A Power BI Compendium

Answers to 65 Commonly Asked Questions on Power BI

Alison Box

Apress®

A Power BI Compendium: Answers to 65 Commonly Asked Questions on Power BI

Alison Box
Billinghurst, West Sussex, UK

ISBN-13 (pbk): 978-1-4842-9764-3   ISBN-13 (electronic): 978-1-4842-9765-0
https://doi.org/10.1007/978-1-4842-9765-0

Managing Director, Apress Media LLC: Welmoed Spahr
Acquisitions Editor: Smriti Srivastava
Development Editor: Laura Berendson
Coordinating Editor: Shaul Elson

Cover designed by eStudioCalamar

Cover image by Sebastian Svenson, unsplash.com

Distributed to the book trade worldwide by Apress Media, LLC, 1 New York Plaza, New York, NY 10004, U.S.A. Phone 1-800-SPRINGER, fax (201) 348-4505, e-mail orders-ny@springer-sbm.com, or visit www.springeronline.com. Apress Media, LLC is a California LLC and the sole member (owner) is Springer Science + Business Media Finance Inc (SSBM Finance Inc). SSBM Finance Inc is a **Delaware** corporation.

For information on translations, please e-mail booktranslations@springernature.com; for reprint, paperback, or audio rights, please e-mail bookpermissions@springernature.com.

Apress titles may be purchased in bulk for academic, corporate, or promotional use. eBook versions and licenses are also available for most titles. For more information, reference our Print and eBook Bulk Sales web page at http://www.apress.com/bulk-sales.

Any source code or other supplementary material referenced by the author in this book is available to readers on GitHub (https://github.com/Apress). For more detailed information, please visit https://www.apress.com/gp/services/source-code.

Paper in this product is recyclable

Table of Contents

About the Author

Alison Box is a Director of Burningsuit Ltd, `www.burningsuit`
`.co.uk`, and an IT trainer and consultant with over 30 years'
experience in delivering computer applications training to all
skill levels, from basic users to advanced technical experts.
She is the author of *Up and Running with DAX for Power BI*
and *Introducing Charticulator for Power BI*, both published by
Apress.

Currently, she specializes in delivering training in
Microsoft Power BI Service and Desktop, Data Modeling,
DAX (Data Analysis Expressions), and Excel. Alison also
works with organizations as a DAX and Data Analysis
consultant. She was one of the first Excel trainers to move into delivering courses in
Power Pivot and DAX, from where Power BI was born. Part of her job entails promoting
Burningsuit as a knowledge base for Power BI and includes writing regular blog posts on
all aspects of Power BI that are published on her website.

About the Technical Reviewer

NS Jenkins, Founder and Director of JPOWER4 and Modern Workplace Solution Architect, is a Microsoft Most Valued Professional (MVP) in M365 Development and an author of *Building Solutions with Microsoft Teams*. He is a Microsoft Certified Trainer (MCT) who has been working on SharePoint for more than 20 years and who loves to learn new things in tech and help others learn them too. He is a certified Microsoft 365 and Power Platform SME, and he is passionate about Microsoft 365 and Power Platform, organizes events, and speaks at events and international conferences, most recently on the topics of Power Platform, Copilot, Microsoft Teams, and SharePoint Framework.

Introduction

In today's data-driven world, the ability to transform raw data into actionable insights is a skill that can set you apart in your career. Microsoft Power BI, with its powerful capabilities for data analysis and visualizations, has become an indispensable tool for professionals and organizations seeking to unlock the full potential of their data. However, as with any powerful tool, there are bound to be questions, challenges, and hurdles along the way.

Welcome to *A Power BI Compendium: Answers to 65 Commonly Asked Questions on Power BI and Power Query*. The Compendium is a comprehensive resource that will enable you to implement more challenging visuals, build better data models, use DAX with more confidence, and execute more complex queries. In doing so, you can find and share those important insights into your data that would be challenging to execute for most Power BI users. It's a "next steps" book of which the objective is to take a Power BI user with limited knowledge and increase their understanding of what can be achieved from using the product and therefore better benefit from its deployment.

The book is a compilation of information from my own blog posts that have answered questions generated by many years' experience in conducting Power BI training courses or working as a Power BI consultant. I have also included questions that have come about from commonly sought-for information from the Power BI Desktop community. They cover a wide and diverse range of topics that many Power BI users often struggle to get to grips with or don't fully understand. Examples of such questions are "How can I plot a dynamic reference line?", "Why is the calculation in the total row wrong?", or "Can I import a Word document?" *The Power BI Compendium* was born out of the need to provide clear and concise answers to such questions.

The reason that such questions are frequently asked is that their answers are not inherent in the Power BI Desktop application but require a deeper understanding of specific visualizations and often advanced knowledge of DAX, data modeling, and Power Query. Not many Power BI users have such eclectic knowledge. Therefore, many users spend time and effort "googling" answers and, even then, may not find suitable solutions or the answers they find are not always the optimal ones. This book is a shortcut for you to find the information you require to execute the reports you've been struggling to build. Many report designers using Power BI have never had formal training in the product and only have a working knowledge at best. They are just ordinary folk who want to get on with their day job but still be able to design more complex reports. In short, people probably just like you.

The format of the book is that 65 questions are posed and categorized into chapters according to their subject matter. The answers are written with the assumption that you have a basic knowledge of Power BI with limited experience in using DAX, Power Query, and building relationships between tables. Each question posed will then take you through the solution in a non-technical and easy-to-follow explanation. Solutions can be adapted or reused in different data analysis contexts, and having the answers to these questions collated in one single book will alleviate the frustrations of Internet searching.

The answers to each question are self-contained, and the order of the questions and chapters is purely arbitrary. Therefore, you can dip in and out and browse the information as you see fit. Feel free to dive into this compendium from cover to cover, or use it as a reference guide, jumping directly to the questions that pertain to your current needs.

There are also companion ".pbix" files that set out the answers to each question, so you can follow along with the examples given in the book. For Chapter 9, "Power Query," the data I have used is contained in Excel files. You can find all these files at `https://github.com/Apress/`. In the "Find a repository..." search box, you can just type "Box."

You will note from the screenshots shown that I have enabled Power BI Desktop's new "On Object Interaction." At the time of writing this feature is currently still in preview. For more information on this preview feature, you can visit:

`https://powerbi.microsoft.com/en-au/blog/on-object-public-preview-opt-in/`

To turn on preview features in Power BI Desktop, on the **File** tab, under **Options and Settings**, select **Options**, and the **Preview features** are under the Global category.

By the time you finish reading this book, you'll not only have answers to 65 common questions but also the confidence to explore, analyze, and visualize data in ways perhaps you never thought possible. Welcome to the future of data-driven decision-making.

CHAPTER 1

Visualizations

On the one hand, Power BI visualizations are easy to generate on the report canvas but on the other hand, they can sometimes leave you feeling frustrated. You select a visual from the gallery that has an approximation to your requirements, add the fields that comprise the visual, and then using the Format pane, work through a myriad of options and settings to arrive at a visual that continues to remain unsatisfactory. Often getting to the visual that truly tells the story of your data can be elusive because many of the features you want to adopt are not found within any of the observable settings of the visual. Here, we address this problem by posing questions where the answer is not intuitive or easy to find, and many of the answers can be re-purposed for different situations and scenarios.

Q1. Can Visual Titles List the Items Filtered in the Visual?

The answer is yes, they can, but someone must inform me how creating such titles became part of the "conditional formatting" options, but let that rest for the moment. There is no "formatting" involved. Instead, you must design DAX measures that generate the text you want to display in the titles of your visuals and then apply them via the conditional formatting settings.

You can see our objective in Figure 1-1 where we've designed a column chart that shows Total Revenue by WINE with a slicer that filters the wines by supplier. This visual has a default title, "Total Sales by WINE." It's in the subtitle that we show the suppliers being filtered. Note that if there are more than two suppliers selected in the slicer, to prevent clutter, the subtitle shows "and more...."

© Alison Box 2023
A. Box, *A Power BI Compendium*, https://doi.org/10.1007/978-1-4842-9765-0_1

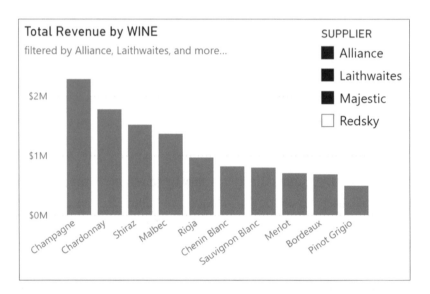

Figure 1-1. *The filtered items are listed in the subtitle of the column chart*

The DAX measure required to generate the subtitle in Figure 1-1 is quite challenging to author, and therefore, we'll take a step-by-step approach. Also, for context, in Figure 1-2, we've shown the Products table that holds the WINE and SUPPLIER columns used in the examples.

WINE ID ▾	WINE ▾	SUPPLIER ▾↑
3	Chardonnay	Alliance
12	Chenin Blanc	Alliance
13	Shiraz	Alliance
14	Lambrusco	Alliance
1	Bordeaux	Laithwaites
2	Champagne	Laithwaites
4	Malbec	Laithwaites
8	Pinot Grigio	Majestic
9	Merlot	Majestic
10	Sauvignon Blanc	Majestic
11	Rioja	Majestic
5	Grenache	Redsky
6	Piesporter	Redsky
7	Chianti	Redsky

Figure 1-2. *The WINE and SUPPLIER columns in the Products table*

Let's begin with a simple measure that will go some way to achieving our desired goal:

```
Title Conditional Formatting #1=
"Filtered by " & SELECTEDVALUE ( Products[SUPPLIER],
"many Suppliers" )
```

This measure uses the SELECTEDVALUE function that returns the value in the column referenced (i.e., Products[SUPPLIER]), but only if *one* value has been filtered in the Products table; otherwise it will return an alternate result (i.e., "many Suppliers").

To apply this subtitle to the visual, in the **Format** pane, under the **Title** card and **Subtitle** subcard, in the **Text** option, click the "*fx*" button, and then use this measure in the "**What field should we base this on?**" dropdown box (Figure 1-3).

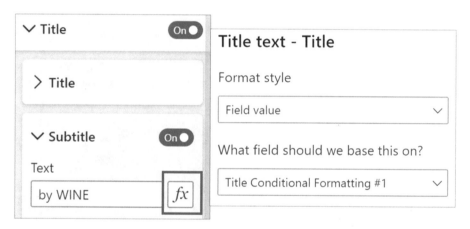

Figure 1-3. *Use the "fx" button on the Subtitle formatting card*

The subtitle generated by this measure is shown in Figure 1-4. It's correct that "Laithwaites" shows in the top chart, but we've still got a problem when no suppliers are selected, as shown in the bottom chart. We haven't filtered "many Suppliers" as stated in the subtitle.

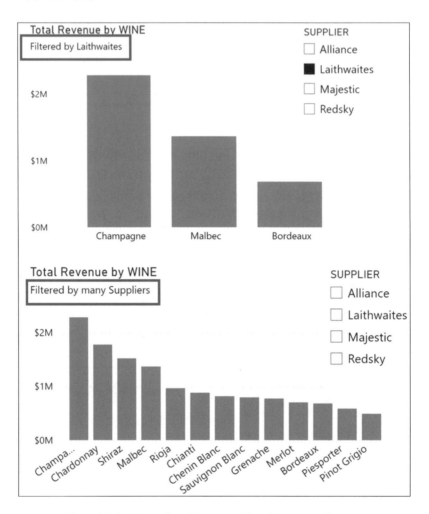

Figure 1-4. *A single selection in the slicer works, but no selection does not*

To resolve this problem, we must be more adventurous with our DAX measure. This measure will list all the values selected in the slicer. However, if there is no selection in the slicer, it will show nothing:

```
Title Conditional Formatting #2 =
VAR NoFilteredSuppliers =
    COUNTROWS ( VALUES ( Products[SUPPLIER] ) )
VAR NoAllSuppliers =
    COUNTROWS ( ALL ( Products[SUPPLIER] ) )
RETURN
    IF (
```

```
NoAllSuppliers = NoFilteredSuppliers,
"",
"filtered by "
    & CONCATENATEX (
        VALUES ( Products[SUPPLIER] ),
        Products[SUPPLIER],
        " & ",
        Products[SUPPLIER],ASC
    )
)
```

This code first checks that the number of suppliers being filtered is equal to the total number of suppliers. If so, no suppliers are filtered (or all suppliers are filtered) and therefore the measure returns nothing. If suppliers are filtered, the CONCATENATEX function is used to list them, separating each supplier name with an ampersand. The last line of the code ensures that the list of suppliers is sorted ascending by supplier name.

Figure 1-5 shows the result of this measure if we place it in the "**What field should we base this on?**" dropdown box of the option of the conditional formatting options.

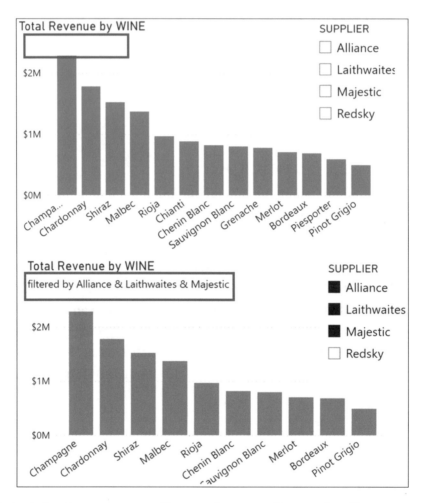

Figure 1-5. *Selecting many suppliers and no suppliers in the slicer*

However, what do you feel about the "&" separator used between the supplier names? Personally, I think it looks a little clumsy, particularly if we had many suppliers that could possibly be selected in the slicer. They would all be separated with an ampersand! Besides, we can improve on this. Let's now design a measure that will account for these three conditions:

1. If two suppliers are selected, use an ampersand.

2. If three suppliers are selected, use a comma.

3. If more than three suppliers are selected, show "and more...."

This is the measure that will do this job:

```
Title Conditional Formatting #3 =
VAR NoFilteredSuppliers =
    COUNTROWS ( VALUES ( Products[SUPPLIER] ) )
VAR NoAllSuppliers =
    COUNTROWS ( ALL ( Products[SUPPLIER] ) )
RETURN
    IF (
        NoAllSuppliers = NoFilteredSuppliers,
        "",
        "filtered by "
            & IF (
                NoFilteredSuppliers > 2,
                CONCATENATEX (
                    TOPN ( 2, VALUES ( Products[SUPPLIER] ),
                    Products[SUPPLIER], ASC ),
                    Products[SUPPLIER],
                    ", ",
                    Products[SUPPLIER], ASC
                ) & ", and more…",
                CONCATENATEX (
                    VALUES ( Products[SUPPLIER] ),
                    Products[SUPPLIER],
                    " & ",
                    Products[SUPPLIER], ASC )))
```

This measure is similar to the previous measure in that it starts by testing if there are no selections in the slicer and if so, returns nothing. If there are more than two suppliers selected, the CONCATENATEX function builds a temporary table that contains just the top two suppliers alphabetically (using the TOPN function) and lists these separated by a comma, concatenating ", and more…." If there are only two suppliers selected, CONCATENATEX will list them separated by an ampersand.

In Figure 1-6, you can see how filtering more than two supplies results in less cluttering of the visual.

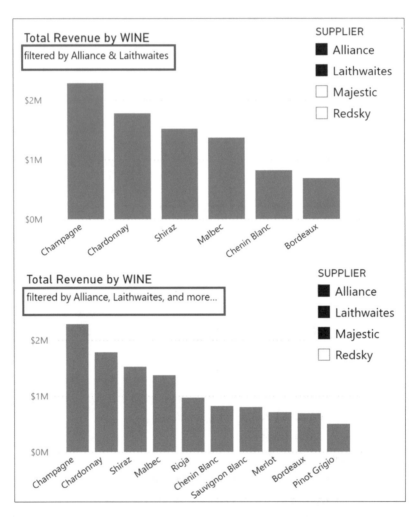

Figure 1-6. *Selecting two suppliers and more than two suppliers in the slicer*

You now know how to list items filtered in slicers in the title or subtitle of visuals. By breaking down the answer into smaller "bite-size" pieces, you understand how the need for a more challenging DAX measure was necessary. You can, of course, use the same measures if you are using the Filters pane to make your selections rather than using slicers.

Q2. Is It Possible to Conditionally Color Markers on Line Chart?

There is a secret to the answer to this question. That is, you start with a *column* chart and apply the conditional colors to the columns. Then, you convert the column chart to a line chart or area chart. For example, have you ever wanted to show high and low points for monthly sales revenue on a line chart, as shown in Figure 1-7?

Figure 1-7. *Conditionally colored markers showing low and high points*

If the answer is yes, then the starting point is to create this DAX measure that assigns the color red to the maximum and minimum values for each month that will be used for the conditional formatting:

```
Max & Min Monthly Sales =
VAR mymax =
    MAXX (
        ALL ( DateTable[Year], DateTable[Month],
            DateTable[Month No.] ),
        [Total Revenue]
    )
VAR mymin =
```

```
    MINX (
        ALL ( DateTable[Year], DateTable[Month],
              DateTable[Month No.] ),
        [Total Revenue]
    )
RETURN
    IF ( [Total Revenue] = mymax ||
        [Total Revenue] = mymin,  "red" )
```

Now create a column chart that contains the same fields that you want in your Line or Area chart. On the **Columns** formatting card and using the conditional formatting "*fx*" button, select "Field value" in the **Format** style dropdown. In "**What field should we base this on?**" dropdown, select the "Max & Min Monthly Sales" measure, as shown in Figure 1-8.

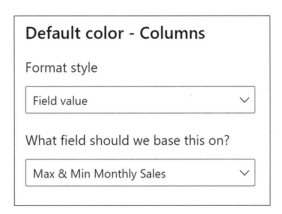

Figure 1-8. *Use the "Field value" Format style to apply the measure*

This will color the column with the lowest and highest values red as shown in Figure 1-9.

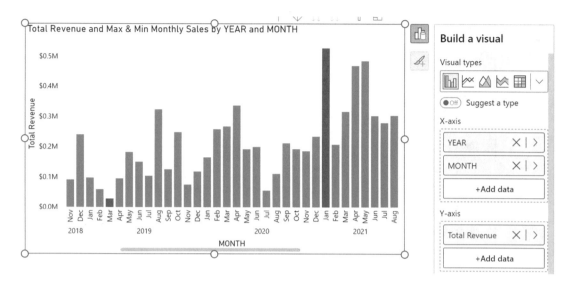

Figure 1-9. *Use the conditional formatting on a column chart to start*

Now simply convert your column chart to a line chart.

Note   In the examples illustrated, we have used the "Smooth" Line Type option on the Lines formatting card.

Interestingly, these markers are not considered as markers on the line chart (on the Format pane, Markers are turned off). However, if you want to add normal markers, use the **Markers** formatting card on the Format pane, turn the markers on, and format accordingly; see Figure 1-10.

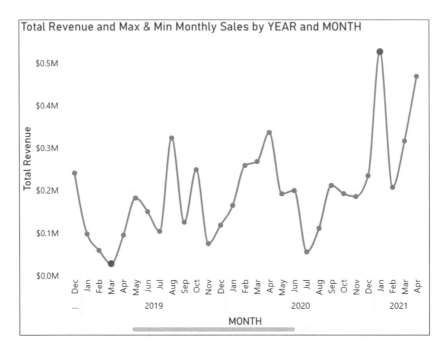

Figure 1-10. *You can add normal markers*

I think you'll agree that finding where you can conditionally format markers is not something you would stumble across in your normal day-to-day interactions with Power BI, but now you've been let into the secret!

Q3. Can Text Boxes Replace Cards by Showing Dynamic Values?

When you want to summarize key metrics in your report, the go-to visualization is the ubiquitous card visual. However, there is an alternative approach to placing a plethora of cards on the canvas. You could use dynamic values in the smart narrative visual or text box instead.

The problem with using cards is that you need a separate card for every metric you want to use in your summary.

Note   There is now a "New Card" visual in preview (at the time of writing) that will dynamically group cards. However, the problem persists that a separate card is generated for each measure.

Take, for example, the visuals in Figure 1-11. Here, we are summarizing sales for salesperson "Abel" for the region "Canada." We have initially done this in six separate cards whereas we could replace all six cards with one visual, as shown in the bottom example. Welcome to the smart narrative visual also known ordinarily as the text box!

Figure 1-11. *Cards can be replaced with a text box*

Although the text box and the smart narrative are essentially the same visual, there is a difference. If you select the smart narrative visual from the Visualizations gallery, Power BI will automatically summarize the data comprising the visuals on the page, whereas if you select a text box, you will need to generate the summary values yourself.

For example, if we insert the smart narrative visual, these are the summaries the visual generates from the existing visualizations; see Figure 1-12.

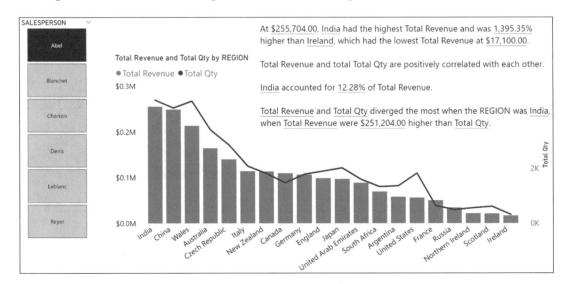

Figure 1-12. *The smart narrative visual summarizes data in the visuals*

The values with the blue underline (the underline doesn't show when you publish) are dynamic values, generated by the visual. If you click one of these values, and then click the **Value** button on the smart narrative menu bar, you can format them if required (Figure 1-13).

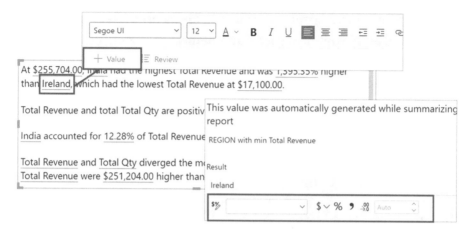

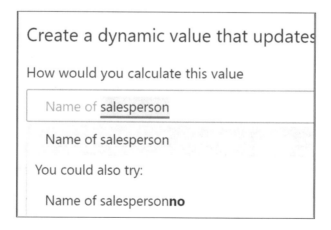

Figure 1-13. *Use the Value button to format values with blue underlines*

You can add your own dynamic values into the smart narrative. To do this, type the text you want, and then to insert a dynamic value, click the **Value** button in the menu as shown in Figure 1-13. In the "**How would you calculate this value**" box, type the value you want to insert. You can use natural language here, just as you would in Q&A. For example, to find the name of the selected salesperson, you could type "Name of salesperson" as shown in Figure 1-14.

Figure 1-14. *You can use natural language to find the value you require*

The blue line under the text you've typed indicates that the smart narrative has recognized the natural language you've used. Click **Save** to insert the value into the smart narrative.

Now you know how to add your own values to the smart narrative; you could use a text box instead and build your own smart summaries.

By "smart" we mean that the summary values will change dynamically as you browse, slice, and filter your report. Use the **Value** button to insert your dynamic values. For example, in Figure 1-15, we've typed "region name" into the "**How would you calculate this value**" box to insert the Region name.

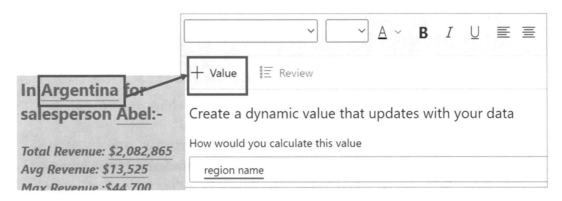

Figure 1-15. *You can insert dynamic values into a text box*

If you want to include the results of DAX measures, you can type their name into this box (excluding the square brackets). If you don't have existing measures, you can ask the text box to do the calculation for you by using natural language, for example, you could type "average revenue" or "sum of revenue"; see Figure 1-16.

Figure 1-16. *You can use natural language for calculations*

This is one of the great benefits of using these dynamic values; you don't always need to know how to do the DAX.

The smart narrative and text box can make great alternatives to using multiple cards. However, we can take this one stage further because the text box can be used instead to supply a necessary narrative.

Take the example in Figure 1-17 where we have filtered "Chardonnay" wine and a start and end range of transaction dates using slicers. We then used a text box to explain the Total Qty values for the selected dates.

WINE	SALE DATE	SALE DATE ▲	Total Qty
☐ Bordeaux	☐ 06 January 2018	06 January 2018	216
☐ Champagne	☐ 17 March 2018	17 March 2018	99
■ Chardonnay	■ 21 April 2018	21 April 2018	201
☐ Chenin Blanc	■ 30 April 2018	30 April 2018	197
☐ Chianti	☐ 18 June 2018	18 June 2018	291
☐ Grenache	☐ 25 June 2018	25 June 2018	217
☐ Lambrusco	☐ 21 July 2018	21 July 2018	295
☐ Malbec	☐ 27 August 2018	27 August 2018	135
☐ Merlot	☐ 12 September 20...	**Total**	**23,737**
☐ Piesporter	☐ 30 March 2019		
☐ Pinot Grigio	☐ 18 April 2019	On 30 April 2018, the Qty for	
☐ Rioja	☐ 10 May 2019	Chardonnay was 197 cases.	
☐ Sauvignon Blanc	☐ 16 August 2019	On 21 April 2018 it was 201	
☐ Shiraz	☐ 22 August 2019		
	☐ 25 September 20...		
	☐ 03 October 2019		

Figure 1-17. *The text box can be used to generate narratives for your data*

We did this by typing natural language into the "**How would you calculate this value**" box as follows:

> "last saledate" returned **30 April 2018**

> "wine name" returned **Chardonnay**

> "last total qty" returned **197**

> "first saledate" returned **21 April 2018**

> "first total qty" returned **201**

If you generated measures for these values, you could also find the difference between the first and last quantities and show this in the text box.

Therefore, the next time you feel the urge to ply cards onto your report page, consider using a single text box instead. But better still, why not generate your own smart narratives that can truly tell the story of your data?

Q4. How Can I Create a Continuous X-Axis on Column Charts for Year and Month Labels?

Have you ever wanted to use a continuous X-axis for month and year labels on a column chart, as shown in Figure 1-18?

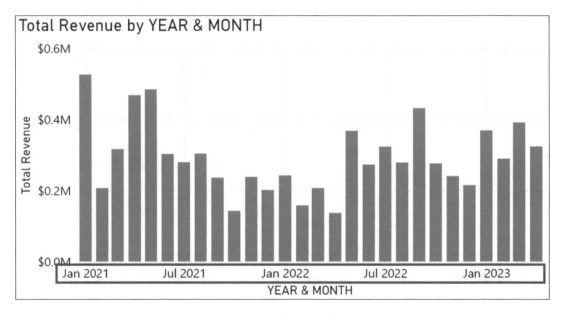

Figure 1-18. *A continuous X-axis on a column chart*

A continuous X-axis shows the year and month labels incrementally spaced according to the number of data points plotted. This is compared to a categorical X-axis where there is a label for every data point.

Luckily, Power BI will provide you with a continuous X-axis if you use a date hierarchy in conjunction with a line chart. Here, you can use the "**Expand down all one level in the hierarchy**" button in the visual header to see data for each level, and the axis labels respond accordingly, generating a continuous axis for each level in the hierarchy as shown in Figure 1-19.

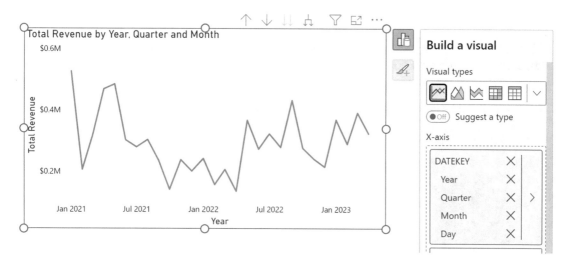

Figure 1-19. *A continuous X-axis in a line chart using a date hierarchy*

However, what if you don't want to plot your data using a line chart? What if you would prefer to use a column chart? If you convert the line chart to a column chart, you are given a categorical X-axis with no ability to change it to continuous, as shown in Figure 1-20.

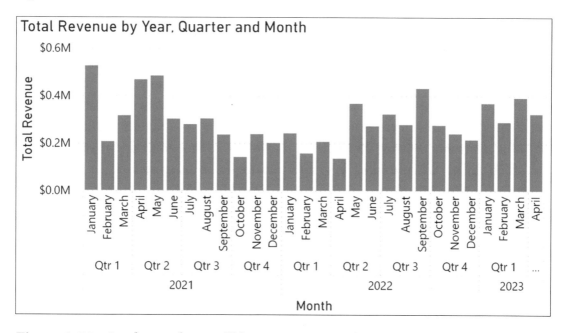

Figure 1-20. *A column chart will have a categorical X-axis*

This is particularly frustrating. The rule is that if you are using a column chart, only numeric data can be plotted on a continuous X-axis. At the month level, the labels comprise both quarter and month names that are text values and so must be plotted on a categorical axis. However, if you drill back up to the year level, you can change the axis from categorical to continuous because year is a numeric value, as shown in Figure 1-21.

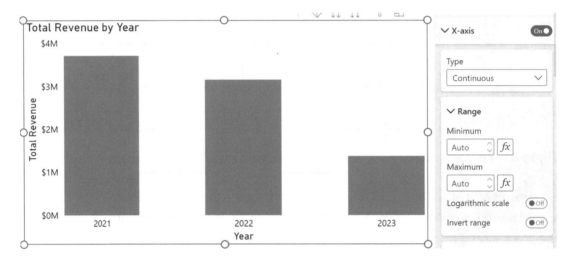

Figure 1-21. *At the year level, you can select a continuous axis*

Now that we have established that with a column chart, you can only use numeric data on the X-axis to render a continuous axis; how can we include both the month name and year in the labels? The answer is that you must generate a column of a date data type (and therefore a numeric value) and format this column to show only the year and month. To do this, create a calculated column in your Date dimension that returns the first of every month using the DAX STARTOFMONTH function. You can then use the **Format** box to format the date as "yyyy mmm" as depicted in Figure 1-22.

File	Home	Help	External tools	Table tools	Column tools

Name: YEAR & MONTH $% Format: yyyy mmm   Σ Summariz

123 Data type: Date/time $ ⌄ % 9 .00 →.0   Auto   Data cate

Structure   Formatting

YEAR & MONTH = STARTOFMONTH(DateTable[DATEKEY])

DATEKEY	YEAR	QTR	MONTH NO.	MONTH	YEAR & MONTH ···
01 July 2018	2018	Qtr 3	7	Jul	2018 Jul
02 July 2018	2018	Qtr 3	7	Jul	2018 Jul
03 July 2018	2018	Qtr 3	7	Jul	2018 Jul
04 July 2018	2018	Qtr 3	7	Jul	2018 Jul
05 July 2018	2018	Qtr 3	7	Jul	2018 Jul
06 July 2018	2018	Qtr 3	7	Jul	2018 Jul
07 July 2018	2018	Qtr 3	7	Jul	2018 Jul
08 July 2018	2018	Qtr 3	7	Jul	2018 Jul
09 July 2018	2018	Qtr 3	7	Jul	2018 Jul
10 July 2018	2018	Qtr 3	7	Jul	2018 Jul
11 July 2018	2018	Qtr 3	7	Jul	2018 Jul
12 July 2018	2018	Qtr 3	7	Jul	2018 Jul
13 July 2018	2018	Qtr 3	7	Jul	2018 Jul
14 July 2018	2018	Qtr 3	7	Jul	2018 Jul
15 July 2018	2018	Qtr 3	7	Jul	2018 Jul

Figure 1-22. *Use STARTOFMONTH to generate a "YEAR & MONTH" column*

You can now use this calculated column on the X-axis of your column chart. However, the default when you place this column in the X-axis bucket is that it will render as a date hierarchy. To prevent this, right-click the YEAR & MONTH field in the X-axis bucket, and select from the shortcut menu accordingly (Figure 1-23).

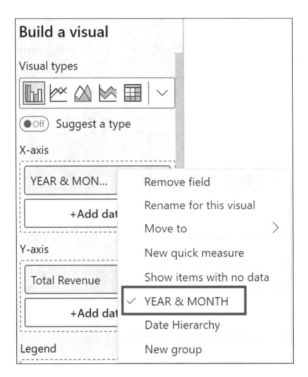

Figure 1-23. *Select YEAR & MONTH, not Date Hierarchy*

Because the YEAR & MONTH column is a date, and therefore a numeric value, it can be plotted on a continuous X-axis.

Q5. How Can I Show Target Lines on Column Charts?

So you want to design a column chart that shows target metric values against the actual metric values? For example, each of our products has a target sales revenue value recorded in the Products table (Figure 1-24), and we would like to visualize this data against their total sales revenue value.

WINE ID	WINE	SUPPLIER	TYPE	WINE COUNTRY	PRICE PER CASE	COST PRICE	TARGET
1	Bordeaux	Laithwaites	Red	France	$75.00	$25.00	£996,531
2	Champagne	Laithwaites	White	Italy	$150.00	$100.00	£2,014,293
3	Chardonnay	Alliance	White	Germany	$100.00	$75.00	£410,527
4	Malbec	Laithwaites	Red	Germany	$85.00	$40.00	£1,750,123
5	Grenache	Redsky	Red	France	$30.00	$10.00	£246,334
13	Shiraz	Alliance	Red	France	$78.00	$30.00	£628,262
6	Piesporter	Redsky	White	France	$135.00	$50.00	£890,463
11	Rioja	Majestic	Red	Italy	$45.00	$15.00	£1,108,976
7	Chianti	Redsky	Red	Italy	$40.00	$10.00	£486,945
12	Chenin Blanc	Alliance	White	France	$50.00	$10.00	£270,565
8	Pinot Grigio	Majestic	White	Italy	$30.00	$5.00	£790,144
9	Merlot	Majestic	Red	France	$39.00	$15.00	£1,014,293
10	Sauvignon Blanc	Majestic	White	Italy	$40.00	$20.00	£589,473
14	Lambrusco	Alliance	White	Italy	$20.00	$15.00	

Figure 1-24. *Target sales revenue values recorded in the Products table*

To accomplish this objective, in Figure 1-25, we have generated two versions of a column chart, both showing the target values as red lines.

Note   If you would like to plot time-based targets, for example, yearly or monthly targets, see Q6 "Can I Create a Constant Line That Responds to Filters?" or Q35 "How Can I Manage Many-to-Many Relationships?"

The top chart was created using the default features in Power BI and is simple to produce, although limited in the options to edit the target lines. The bottom chart was designed using the custom visual, Charticulator. Using Charticulator provides a plethora of design options but is more challenging to generate if you have little or no knowledge of Charticulator.

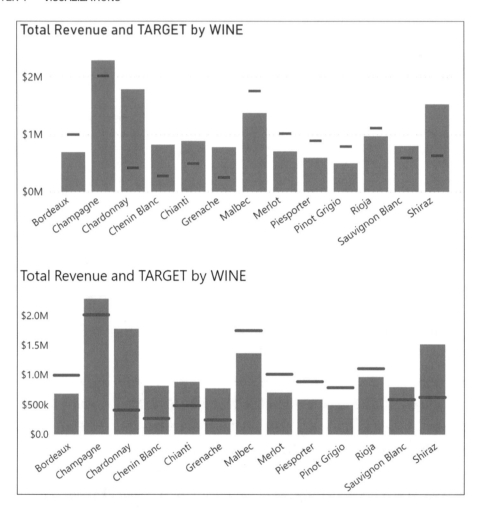

Figure 1-25. *Two versions of a column chart showing the target lines*

Let's first explore how to create the simpler chart. This uses a default visual of Power BI, the line and stacked column chart. Place the TARGET column in the **Line y-axis** bucket; see Figure 1-26.

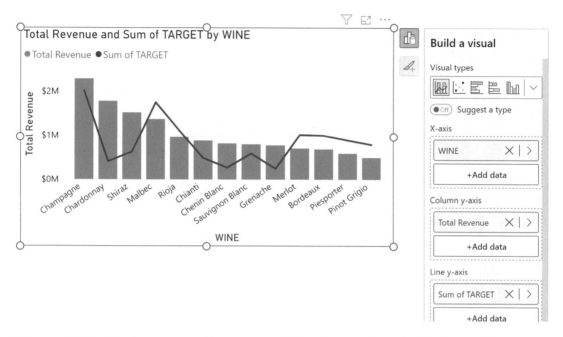

Figure 1-26. *To show target lines, start with a line and stacked column chart*

On the **Format** pane, turn **Markers** on, and in the **Shape** card, change the **Type** option to a line. These will be our target lines, so you may also want to increase the size and change the color. We only want to show the markers, so to remove the line from the chart, in the **Lines** formatting card, edit the **Stroke** option to a width of "0"; see Figure 1-27. This will create a column chart with target lines, generated from line markers.

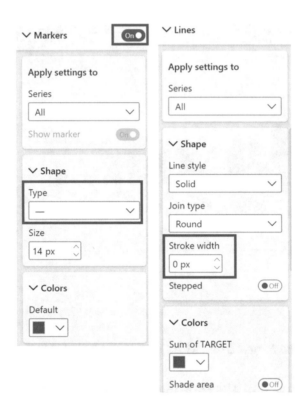

Figure 1-27. *Turn the Markers on and use the line Shape Type. Set the Stroke width of the line to "0."*

The bottom chart in Figure 1-25 was created using Charticulator. Charticulator allows you to generate your very own specialist visuals and graphics. It's a custom visual that can be downloaded into Power BI but is also an application in its own right with its own user interface.

Although it's beyond the remit of this book to explain the workings of Charticulator in detail, I would like to introduce you to one of its more powerful tools, the Data Axis, and take you through an example of how you can use Charticulator to generate your own visualizations.

Note   If you would like to learn more about Charticulator, you can read my book *Introducing Charticulator for Power BI*, published by Apress and available from Amazon.

To download Charticulator, on the **More Visuals** button on the **Home** tab, select "From AppSource" (Figure 1-28).

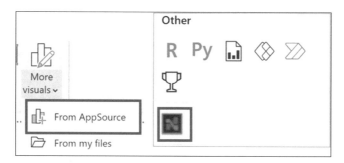

Figure 1-28. *Download Charticulator from the AppSource*

Once downloaded, select the Charticulator visual from the visualizations gallery, and add the fields that will comprise the visual into the Data bucket. In our example, these are WINE, TARGET, and Total Revenue. Then click the **More options** button top right of the visual, select **Edit** and then click on **Create a chart**. This will open Charticulator; see Figure 1-29.

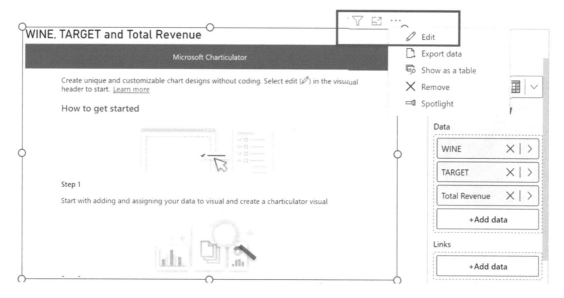

Figure 1-29. *Use the Edit button to open Charticulator*

Once inside Charticulator, from the menu bar, drag the **Data Axis** button into the Glyph pane. Then drag TARGET and Total Revenue (the values to be plotted) from the **Fields** pane onto the Data Axis in the **Glyph** pane; see Figure 1-30.

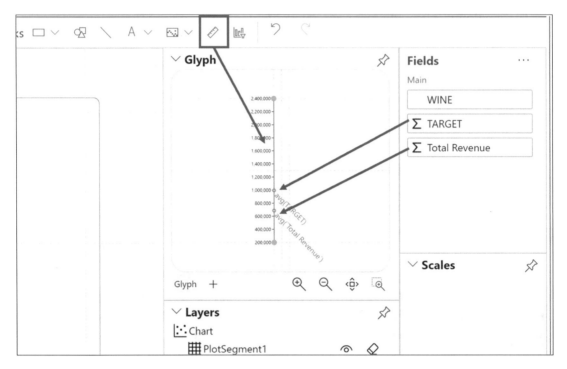

Figure 1-30. *Drag the Data Axis button into the Glyph pane, and drag the values onto the Data Axis*

To create a column, click the **Marks** button, and draw a rectangle aligned to the "avg(Total Revenue)" mark on the Data Axis and within the gray dotted lines. To create the target line, using the **Line** button on the menu bar, draw a line aligned to the "avg(TARGET)" mark on the Data Axis (Figure 1-31).

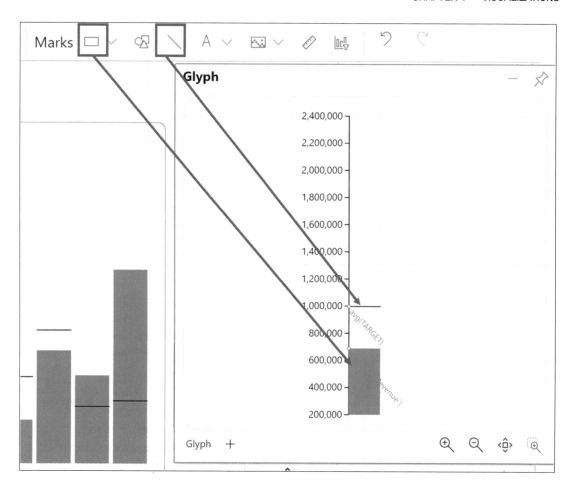

Figure 1-31. *Draw a rectangle and a line aligned with the marks on the Data Axis*

To show the product names on the X-axis, drag the product name field (e.g., WINE) onto the X-axis on Charticulator's chart canvas (Figure 1-32). You may also need to drag on the gray guide lines to pull the X- and Y-axes onto the canvas, shown by red ovals in Figure 1-32.

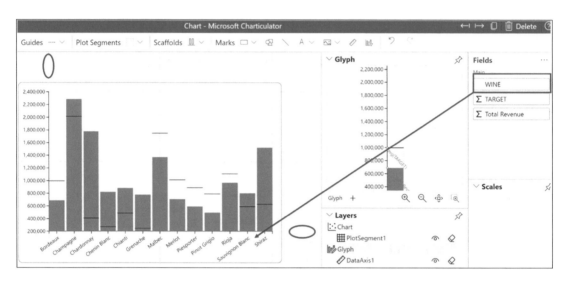

Figure 1-32. *Drag the product name onto the X-axis*

To format the target lines, select "Line1" in the **Layers** pane, and in the **Attributes** pane, you will find options to change the color and width (Figure 1-33). When you have finished, click **Save** and then **Back to report** at the top left of the application.

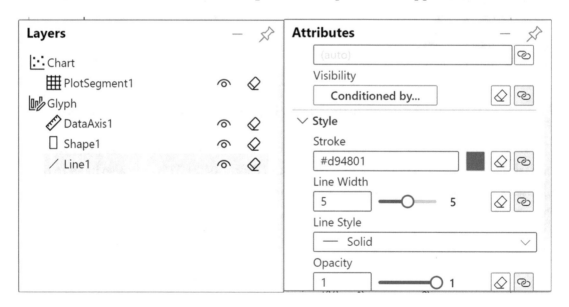

Figure 1-33. *Formatting options are in the Attributes pane*

So now you have a choice as to whether you want to use Power BI or Charticulator to generate target lines on column charts. The benefit of using Charticulator is that in the Attributes pane, there are more formatting options than if you are using markers in the Power BI chart. But also, I hope you enjoyed your first foray into the world of Charticulator. We will be meeting Charticulator's Data Axis again in the next chapter when we answer questions on filtering data.

Q6. Can I Create a Constant Line That Responds to Filters?

Since the launch of Power BI eight years ago, this is a question I've consistently been asked. The "constant" line to which people are referring is one of the reference line types found on the **Reference line** formatting card (see Figure 1-34).

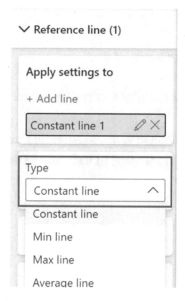

Figure 1-34. *The "Constant" reference line*

Until the February 2022 release of Power BI, the answer was always no, but now the great news is that you can add constant lines ***that are not constant but change with user selections*** from slicers or from the Filters pane.

For instance, in your data model, you may have a table that records the monthly sales targets for your salespeople as depicted in Figure 1-35.

YEAR ▾	MONTH ▾	TARGET ▾
2018	Jan	$187,219
2018	Feb	$133,086
2018	Mar	$132,004
2018	Apr	$118,300
2018	May	$144,290
2018	Jun	$167,533
2018	Jul	$109,990
2018	Aug	$107,713
2018	Sep	$132,266
2018	Oct	$100,418
2018	Nov	$129,281
2018	Dec	$175,014
2019	Jan	$137,587
2019	Feb	$124,441

Figure 1-35. *Monthly Targets table*

Please note that this Targets table would not be related to any other tables in your data model.

Ideally, you would then like to generate a dynamic line that shows these Targets according to the year and month selected in a slicer, as shown in Figure 1-36 where the target for January 2023 is **$78,085** and for February 2023 it is **$71,523**.

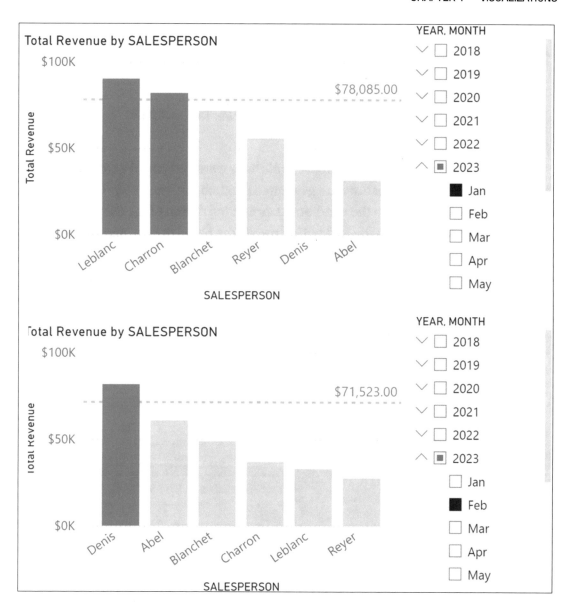

Figure 1-36. *A Constant line that changes with the month selected in the slicer*

In the visuals in Figure 1-36, we have also used conditional formatting on the columns so that salespeople whose sales are below the target are colored gray.

If you would like to generate similar visuals using the dynamic Constant line and conditional formatting, the starting point is to create a DAX measure that can be used for the Constant line value. The Constant line must respond to the filtering of the year and month in the slicer, and therefore the measure must filter the relevant year in the "Targets" table accordingly:

```
Target =
CALCULATE (
    SUM ( Targets[TARGET] ),
    Targets[YEAR] = SELECTEDVALUE ( DateTable[YEAR] )
        && Targets[MONTH] = SELECTEDVALUE ( DateTable[MONTH] )
)
```

Just as in the answer to QU1, where dynamic titles were associated with conditional formatting, dynamic Constant lines are generated in much the same way. Therefore, you must now add this measure to the conditional formatting option of the Constant line.

On the **Line** card, click on the "*fx*" button beside the **Value** option, and in the conditional formatting dialog, select "**Field value**" as the Format style. Then select your measure in "**What field should we base this on?**" dropdown as shown in Figure 1-37.

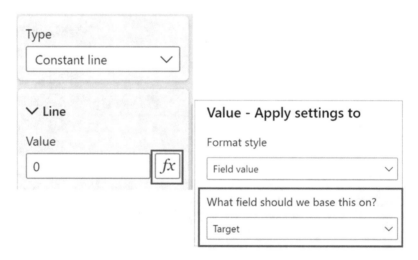

Figure 1-37. *Adding the measure to the Constant line's conditional formatting*

To apply the conditional formatting to the columns or bars of your chart, create another measure that will attribute a color to the columns or bars below the target. For example:

```
Conditional Format Target =
IF( [Target] > [Total Revenue], "Light Grey" )
```

You can use the "*fx*" button beside the default color option on the **Columns** formatting card and select this measure in "**on which field should we base this on?**"; see Figure 1-38.

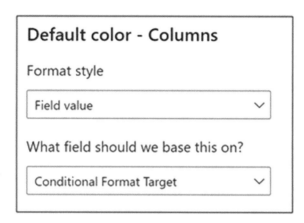

Figure 1-38. *Applying conditional formatting to the columns of the chart*

So, at long last, after eight years, we have finally arrived at a user-defined reference line that does what we want it to and that is to respond to values being filtered in the visual.

Q7. How Do I Create a Waterfall Chart with "From" and "To" Values?

If you have ever created the default waterfall visual in Power BI, you will know that it will only show you the "To" value. This value is the total of the individual values from each category. For example, in Figure 1-39, the "Total" column is simply the sum of yearly revenues.

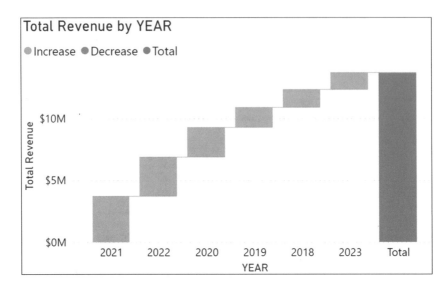

Figure 1-39. *The default Power BI waterfall visual*

In fact, the only way you can get "From" and "To" totals is to introduce a "Breakdown" field. For example, in Figure 1-40, the "REGION" field has been placed in the "Breakdown" bucket, and we're also only considering the years 2021 and 2022.

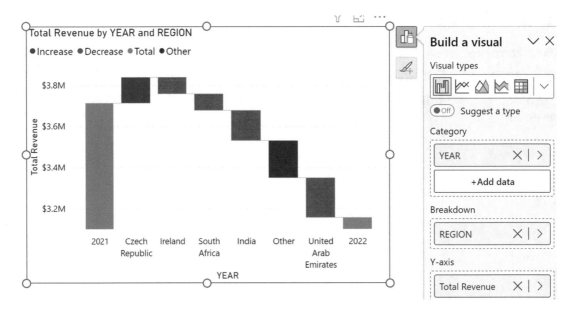

Figure 1-40. *Introducing a "breakdown" renders "To" and "From" totals*

But what if you want to plot "From" and "To" values that are calculated by measures and then show the breakdown variances, for example, the cost variances between Turnover and Profit? See Figure 1-41.

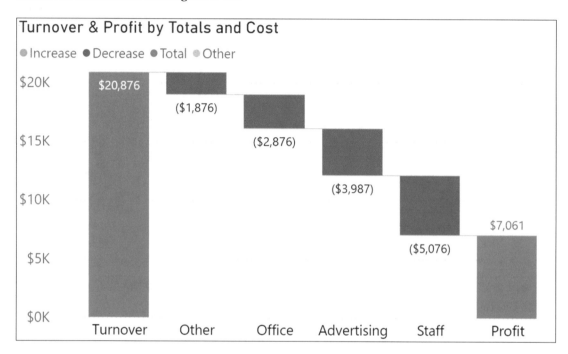

Figure 1-41. *Breakdown variances for Turnover and Profit*

In Figure 1-42 we illustrate another example where, rather than static years plotted, we have plotted the previous year's sales (determined by the slicer selection) and show the breakdown of our salespeople's sales variances.

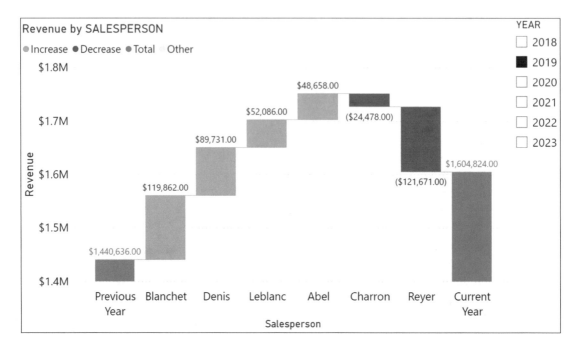

Figure 1-42. *Salespeople's breakdown from Previous Year (2018) to Current Year (2019)*

So how were these waterfall charts generated? Let's first take the "Turnover & Profit" example in Figure 1-41.

First, you must build the DAX measures that calculate the values to be plotted in the waterfall chart. For example, Turnover Total, Profit Total, and each individual total cost (e.g., Total advertising, Total staff, Total office, etc.).

The next requirement is to generate tables that contain the labels for the columns in the waterfall chart. You can use the **Enter data** button on the **Home** tab to do this. There are two sets of labels required; one set is for the "Main Category," and the other set is for the "Breakdown." Figure 1-43 shows the two tables that are required.

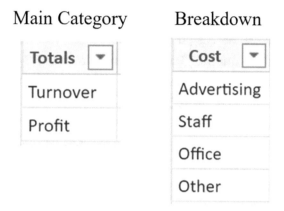

Figure 1-43. *The two tables required for the labels in the "Turnover & Profit" Waterfall*

Finally, we can now create this measure that will be used by the visual:

```
Turnover & Profit =
SWITCH (
    SELECTEDVALUE ( 'Main Category'[Totals]),
    "turnover",
        SWITCH (
            SELECTEDVALUE ( Breakdown[Cost] ),
            "advertising",[Total advertising],
            "staff",[Total staff],
            "office",[Total office],
            "other",[Total other],
            [Turnover Total]
        ),
    "profit",
        SWITCH (
            SELECTEDVALUE ( Breakdown[Cost] ),
            "advertising",0,
            "staff",0,
            "office",0,
            "other",0,
            [Profit Total]))
```

The measure works by subtracting the "To" column values (i.e., the profit values) from the "From" column values (i.e., the turnover values) to generate the breakdown values. We are now ready to put the "Totals" column from the "Main Category" table into the Category bucket of the waterfall chart and the "Cost" column from the "Breakdown" table into the Breakdown bucket. Into the Y-axis, of course, goes the measure. You must also ensure that the visual is sorted by "Totals" descending.

For the second example in Figure 1-42 where we show the salespeople's revenue breakdown from a previous year to a current year, we only need to create one additional table that will label the category columns in the waterfall. We called this table "Categories"; see Figure 1-44.

Figure 1-44. *The table required for the column labels in the "Revenue by Salesperson" Waterfall*

We already have a Salespeople dimension where the names of each salesperson are recorded in the SALESPERSON column, so we don't need a table to generate these labels for the breakdown. We also have a Total Revenue measure that calculates the current year's revenue. However, we do need a measure that calculates the previous year's revenue from the year selected in the slicer:

```
Previous Yr =
CALCULATE([Total Revenue],DATEADD(DateTable[DateKey],-1,YEAR))
```

This is the measure that will drive the waterfall visual in Figure 1-42:

```
Revenue =
SWITCH (
    SELECTEDVALUE ( Category[Category Value]),
    "Previous Year",
        SWITCH (
            SELECTEDVALUE ( SalesPeople[SALESPERSON] ),
```

```
            "reyer", CALCULATE([Previous Yr],
                SalesPeople[SALESPERSON]="reyer"),
            "denis", CALCULATE([Previous Yr],
                SalesPeople[SALESPERSON]="denis"),
            "blanchet", CALCULATE([Previous Yr],
                SalesPeople[SALESPERSON]="blanchet"),
            "leblanc", CALCULATE([Previous Yr],
                SalesPeople[SALESPERSON]="leblanc"),
            "charron", CALCULATE([Previous Yr],
                SalesPeople[SALESPERSON]="charron"),
            [Previous Yr]
        ),
    "Current Year",
        SWITCH (
            SELECTEDVALUE ( SalesPeople[SALESPERSON] ),
            "reyer", CALCULATE([Total Revenue],
                SalesPeople[SALESPERSON]="reyer"),
            "denis", CALCULATE([Total Revenue],
                SalesPeople[SALESPERSON]="denis"),
            "blanchet", CALCULATE([Total Revenue],
                SalesPeople[SALESPERSON]="blanchet"),
            "leblanc", CALCULATE([Total Revenue],
                SalesPeople[SALESPERSON]="leblanc"),
            "charron", CALCULATE([Total Revenue],
                SalesPeople[SALESPERSON]="charron"),
            [Total Revenue]))
```

Just as before, this measure works by subtracting the "From" column values (i.e.,
the previous year's values) from the "To" column values (i.e., the current year's values)
to generate the breakdown values. To construct the Waterfall chart, you can place
the "Category" column from the "Categories" table into the Category bucket, the
SALESPERSON column into the Breakdown bucket, and the Revenue measure into the
Y-axis bucket; see Figure 1-45.

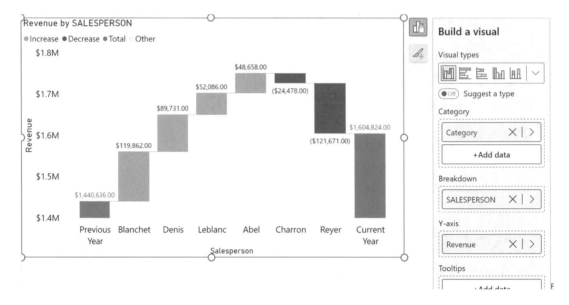

Figure 1-45. *How the Waterfall is constructed*

Therefore, to design your own customized waterfall chart, you must subtract the "From" column values from the "To" column values to generate the breakdown values. Understanding this calculation means that you can design waterfall charts that do what you require of them, which is to show variances from a starting position to an ending one.

Q8. How Do I Show Percentages in a Bar or Column Chart?

The easiest way to show percentage breakdown is in a pie chart. This is because by using the Detail labels, no calculations are required on your behalf (Figure 1-46).

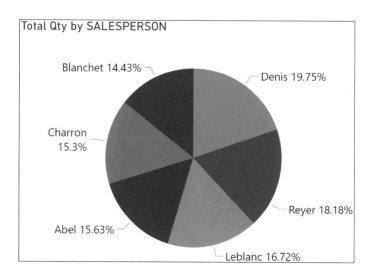

Figure 1-46. *A pie chart will show percentages in the Detail labels*

However, the consensus among data analysts is that pie charts should be used with restraint. This is because it's often difficult to make sense of the information they depict. For instance, in Figure 1-46, how easy is it to see which salesperson has achieved the greatest Total Qty? I believe pie charts are only effective when plotting the ratio between *two* categories, as this is the only scenario where they are easy to decode.

However, how do you get the percentages displayed in data labels if you decide to use some other visual, such as a column chart, bar chart, or a table? The answer to this question depends on what percentage you want shown: percentage of the grand total or percentage of a total that's been filtered. If it's the former, that is, the percentage of the grand total, once you've created your bar chart, you can right-click the value in the X-axis bucket, and select "**Show value as**" and then "**Percent of grand total**"; see Figure 1-47.

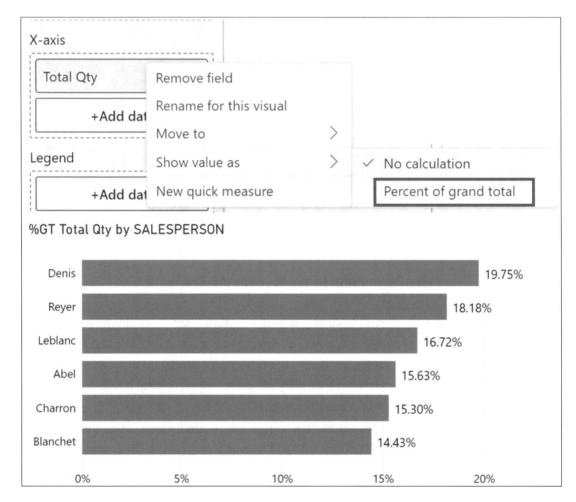

Figure 1-47. *Right-click the value to show percent of grand total*

But how will this percentage respond when filters are applied to the visual? In Figure 1-48 you can see that the percentage is no longer of the grand total, but instead, both the bar and pie charts now show percentages of 2022, filtered in the slicer.

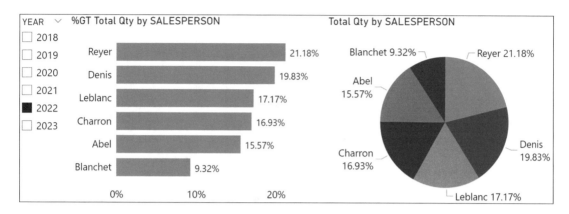

Figure 1-48. *Percentages are for the year 2022, not the grand total*

What's interesting is, if instead of filtering 2022 in a slicer, we highlight this same year using a column chart, the bar and pie charts return different results in the data and detail labels (see Figure 1-49). The pie chart now displays the percent of the grand total, but the bar chart persists in showing percent of 2022.

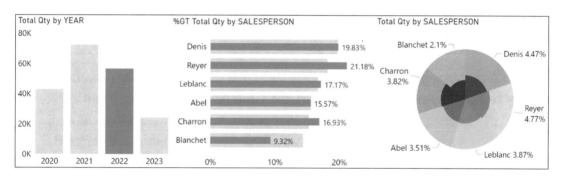

Figure 1-49. *Highlighting 2022 returns different results in the bar and pie charts*

What we can deduce from these examples is that if you want to correctly show percent of the *grand total* in any visual other than a pie chart, you must generate the calculation yourself. Here is the DAX measure you need:

```
%GT Total Qty Measure =
DIVIDE ( [Total Qty], CALCULATE ( [Total Qty], ALL ( Sales ) ) )
```

If you place this measure in the X-axis bucket of the bar chart, you will now see percent of the grand total (see Figure 1-50).

Figure 1-50. *Using the "%GT Total Qty Measure" returns the correct data labels for the bar chart*

In both the bar chart and pie chart, we can now show the percentage of the grand total in the data labels or detail labels, respectively.

Note   Pie charts will only show a percent of the grand total if the interaction is a highlight. A filter interaction will always show the percentage for filter items irrespective of the measure being used.

Let's now move on to tackle another scenario pertaining to showing percentages. In a stacked bar or a stacked column chart, there is often a requirement to show the percentage that each stacked bar or column represents of the whole bar or column. However, if you attempt to show your values as a percent of the grand total using the "Show value as" option and express this in the data labels, clearly this is not what we want; see Figure 1-51.

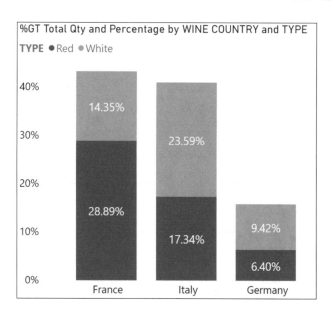

Figure 1-51. *Data labels showing percent of grand total are not what you want*

The solution is again to use DAX to calculate the percentage required:

```
Percentage =
[Total Qty] / CALCULATE ( [Total Qty], ALL ( Products[TYPE] ) )
```

In this measure, in the brackets of the ALL function, you must reference the column that you will use on the X-axis, for example, the TYPE column in our example.

You can now use this measure as a custom data label, ensuring that the display units are set to "None"; see Figure 1-52.

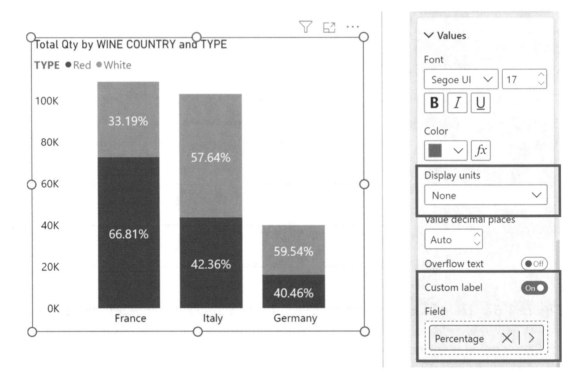

Figure 1-52. *Using a DAX measure in a Custom data label shows the correct percentage*

So if you want to express percentages, you are now no longer a prisoner of the pie chart but can instead use a variety of other visuals. These DAX measures, for example, will work equally well in a table or a matrix visual.

Q9. How Can I Identify the Key Influencers of Important Metrics?

The key influencers visual, introduced in February 2019, was Microsoft's first visual that used "machine learning" to identify factors that influence the outcome of a particular metric.

On the one hand, it's a very simple visual to construct (it only has three buckets: "Analyze," "Explain by," and "Expand by"), but on the other hand, you might struggle to understand the plethora of percentages and increase or decrease factors that it returns. What makes the analysis shown in this visual challenging to fathom is that you will obtain different outcomes depending on whether you're using:

- The "Continuous" analysis

- The "Categorical" analysis

- Analyzing DAX measures (as opposed to data in columns)

We'll be exploring all three aspects of the visual, and the starting point is to use real-life data. After all, what's the point in finding "insights" and "answers" when the data is fictitious?

For this reason, in the examples that follow, we'll be using the Power BI report called "Strictly" which analyzes the data from *Strictly Come Dancing*, the long-running television dance competition.

Note   The BBC *Strictly Come Dancing* television program is a dance competition that, at the time of writing, has completed 20 series with a total of 279 contestants who have danced their way through 2,347 dances and been given 9,388 scores from four judges. For more information visit `www.bbc.co.uk/programmes/b006m8dq`.

Using this data, we would like to identify factors that most influence contestants to achieve better scores for their dances. We all know that some lucky people are just born more talented than others, but above and beyond this, are there any other factors, other than talent alone, that contribute to contestants getting higher scores? For instance, does a particular dance, a particular judge, their age or gender, or even the order in which they dance make a difference to the outcome of their scores? Let's find answers to these questions by using the key influencers visual that you will find in the Visualizations gallery under the "AI visuals" category; see Figure 1-53.

Figure 1-53. *The Key influencers visual*

In the data model of the "Strictly" report, the data we'll be using for the analysis resides in the Scores fact table. In the column called "Score" is recorded each of the four judges' scores, out of 10, for every dance (the maximum score for any dance is therefore 40); see Figure 1-54.

Dance Key ↑	Contestant	Dance	Judge	Score
Abbey Clancy American Smooth 12	Abbey Clancy	American Smooth	Len	10
Abbey Clancy American Smooth 12	Abbey Clancy	American Smooth	Craig	9
Abbey Clancy American Smooth 12	Abbey Clancy	American Smooth	Bruno	10
Abbey Clancy American Smooth 12	Abbey Clancy	American Smooth	Darcey	10
Abbey Clancy Cha Cha Cha 2	Abbey Clancy	Cha Cha Cha	Len	8
Abbey Clancy Cha Cha Cha 2	Abbey Clancy	Cha Cha Cha	Craig	7
Abbey Clancy Cha Cha Cha 2	Abbey Clancy	Cha Cha Cha	Bruno	8
Abbey Clancy Cha Cha Cha 2	Abbey Clancy	Cha Cha Cha	Darcey	7
Abbey Clancy Charleston 7	Abbey Clancy	Charleston	Craig	9
Abbey Clancy Charleston 7	Abbey Clancy	Charleston	Bruno	9
Abbey Clancy Charleston 7	Abbey Clancy	Charleston	Len	9
Abbey Clancy Charleston 7	Abbey Clancy	Charleston	Darcey	9
Abbey Clancy Foxtrot 5	Abbey Clancy	Foxtrot	Len	9

Figure 1-54. *The Score column in the Scores fact table*

Therefore, into the "Analyze" bucket of the key inflluencers visual, we will put the "Score" column. Into the "Explain by" bucket, we will put any factors that we *think* may influence the scores (see Figure 1-55). These will be

- The "Gender" column from the Contestants dimension (the Contestants table is related to the Scores fact table in a many-to-one relationship).

- The "Age" column from the Contestants dimension.

- The "Dance" from the Dances dimension (also related to the Scores fact table).

- The "First or Last column" of the Dances dimension. This records the order they performed each dance, that is, "first", "last", or "middle".

As we place these fields into the "Explain by" bucket, the key influencers visual comes to life and returns answers to the question, "What influences Score to increase?"

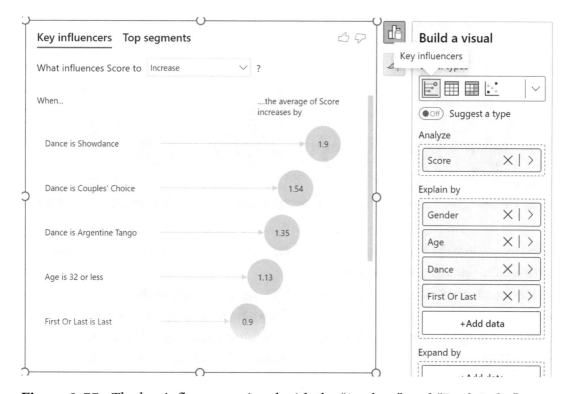

Figure 1-55. *The key influencers visual with the "Analyze" and "Explain by" buckets*

Before we look more closely at these answers, let's first note four important factors:

1. That "Score" is a *column* and not a *measure*.

2. We're using the "Continuous" analysis which is the default if you're using a numeric column in the "Analyze" bucket (Figure 1-56).

3. In the "Analyze" bucket we're using a value from the fact table.

4. There is no summarization on the Score column. The Score field in the "Analyze" bucket does not default to "Sum" but to "Don't summarize" (Figure 1-56).

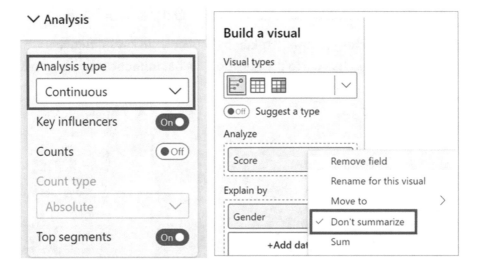

Figure 1-56. *The "Continuous" Analysis type and the "Don't summarize" option*

As we explore the key influencers visual, we will discover how these four factors influence the outcome of the analysis.

Having placed fields in the "Analyze" and "Explain by" buckets, the visual now returns its conclusions; see Figure 1-57. The key influencers visual is divided into two sections, "Key influencers" and "Top segments" (shown at the top left of the visual), so we will need to look at both these areas of analysis. We will start with the "Key influencers" section, where clicking on the bubbles displays a visual on the right (in this instance a column chart) explaining the analysis.

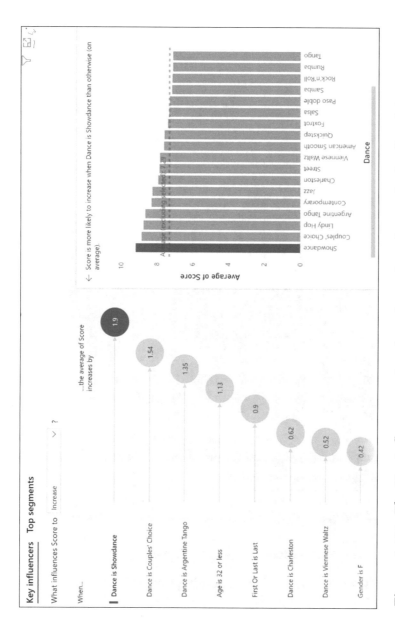

Figure 1-57. *The Key influencers visual analyzes the columns in the "Explain by" buckets*

The visual tells us that the greatest influencer on the score increasing is the "Showdance." If contestants perform this dance, the average of their scores increases by 1.9 (shown in the bubble). To help you compare the key influencer with the other dances, their average scores are shown in a column chart on the right. This chart also tells us that the average for all the other dances is 7.29 shown in the reference line on the column chart; see Figure 1-58.

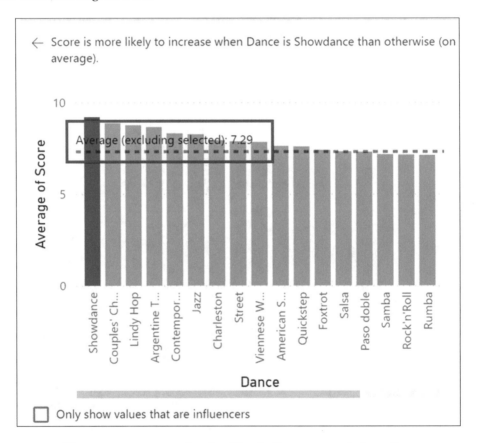

Figure 1-58. *The average scores for the Key influencer and the others*

Other influencers on scores increasing are as follows:

- Dancing "Couples' Choice" or "Argentine Tango"

- Being aged 32 or less

- Dancing last

We now know that if our contestants dance a "Showdance," they should get a better score. However, this raises another question: How likely is it that contestants will get a chance to dance a "Showdance"? They also get a better score (by almost one whole mark) if they dance last, but how likely is it that they will get to dance last? Dancing a "Showdance" or dancing last is not a consistent factor throughout each week of the competition. However, being female or being relatively young is consistent. To throw more light on this aspect of the analysis, we need a way of finding the percentages that these "influencers" comprise our total dataset. What we can do here is ask to **"Enable counts"** on the **Analysis** formatting card and then sort the bubbles by count; see Figure 1-59.

Figure 1-59. *Enabling and sorting by "Counts"*

When we do this, around each bubble, a gray line indicates the percentage of the entire number of scores that this influencer comprises. "Gender is F," which was considered the least of the influencers (scores will only increase by 0.42 if you are female), is now at the top, while "Showdance" is now second from bottom. If you hover over the bubbles, you can see the precise percentages which explain why the influencers are now sorted differently. Scores for female contestants comprise 52.19%, whereas the scores for a "Showdance" comprise only 2.13% of the data. Being female is considered to have a greater influence on higher scores because their scores comprise more than half of all scores. Dancing a "Showdance" comprises too little of the data to be considered significant.

Up till now, we've been examining each individual influencer. However, would it be true to say that if you are young and female, danced last, and danced a "Showdance," you will get a higher average score? This is where "Top segments" comes in. Top segments looks at *all* the key influencers that have been identified and finds out if a particular combination of these leads to higher scores.

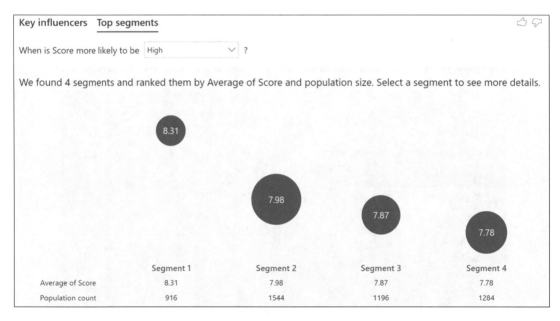

Figure 1-60. *The Top segments of the key influencers visual*

Figure 1-60 shows four segments, represented by the bubbles. Each segment contains a group of contestants that have achieved higher average scores. The bigger the bubble, the more judges' scores are in that segment. Remember that the analysis is being done on the rows in the Scores table, and therefore the "Population count" at the bottom references the number of rows in this table, that is, the number of individual judge's scores. To see the makeup of each segment, click the bubble. Each segment compares the average score of contestants in the segment to the average score overall. Let's explore Segment 1 where contestants get an average score of 8.31, compared to 7.34 overall; see Figure 1-61.

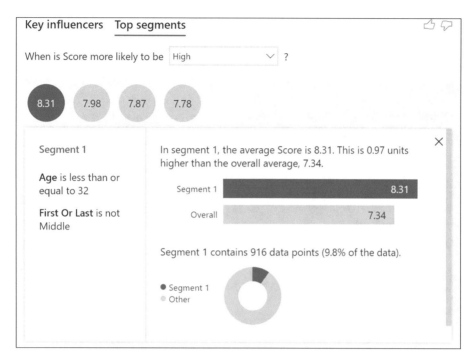

Figure 1-61. *Segment 1 in the Top Segments of the key influencers visual*

Segment 1 contains the scores of contestants who are aged 32 or less and who danced either first or last. Note that we are also shown the percentage that these scores comprise of the whole dataset (9.8%) in a donut.

If we now examine Segments 2, 3, and 4, we will find that nowhere are any of the dances listed as influencing the scores. The "Counts" view shows us why by sorting the dances at the bottom of the list. Contestants only perform each dance once or twice during the competition, so a particular dance, when combined with the other factors (i.e., gender, age, and order), becomes too small a percentage of the overall data to be considered in the segment analysis.

Because we now know that the dance the contestants performed should not be considered as an influencer, let's remove the Dances column from the "Explain by" bucket.

At this point, we need to remind ourselves that, so far, we've been using the "Continuous" type of analysis which means we've been looking at the factors that cause the scores to increase, being measured by calculating averages for higher-scoring groups. But there is another way to look at our contestants' performance. What if we want to find out what factors cause our contestants to get a maximum score of 10? This is where "Categorical" analysis will be useful. Changing to this **Analysis type** by using the option on the **Analysis** card means we can then select the "10" category from the dropdown; see Figure 1-62.

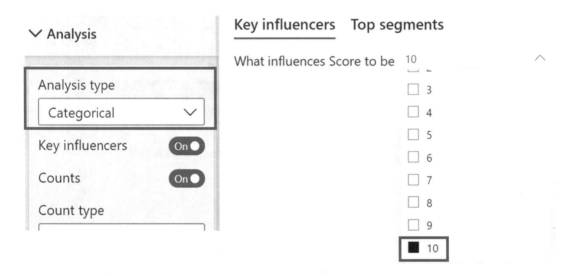

Figure 1-62. *Using the "Categorical" analysis type*

We can see the categorical analysis that shows the key influencers for a score of 10 in Figure 1-63.

Note "Categorical" analysis will work for our data because we have 10 distinct scores. If we had many scores, for example, 1 to 100, then we would need to categorize them accordingly, for instance, into "Very High," "High," "Middle," etc.

Notice that the analysis now uses factors (as opposed to increased values) to determine the extent of influence of a particular group. It tells us that if contestants are aged 32 or less, it's more than twice as likely (2.37 times) that these contestants will score 10, or if you dance last, your chances of getting a 10 are 1.88 times more likely.

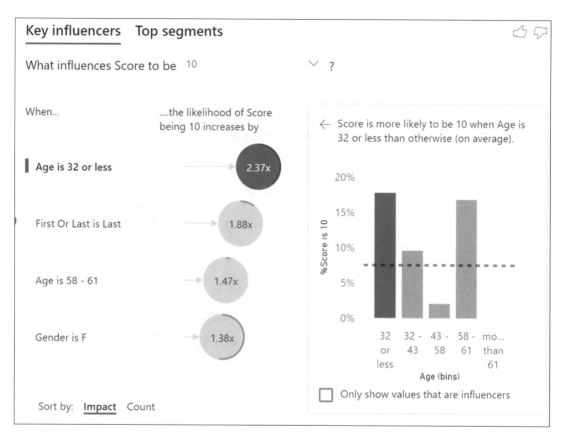

Figure 1-63. *The influencers for a Score of 10*

Use a Categorical analysis if your metrics contain only a few distinct values or if you have fields in your data that define text categories (e.g., "High," "Medium," "Low"). Use the "Continuous" analysis for numeric metrics that have a high cardinality.

You may feel that you would like to use a DAX measure in the "Analyze" bucket. Be warned because a measure will group and aggregate data, and therefore the key influencers visual will be analyzing values at a higher granularity than using the row-level data in the columns of a table that haven't been summarized. This can return unreliable results. The "Expand by" bucket goes some way to resolving this dilemma. You can increase the granularity of the data being used by the measure by adding a row identifier into this bucket such as a primary key, but the key influencers visual works best when you don't use measures.

Q10. Is It Possible to Multi-select with Visual Interactions?

We all know that by default, if we click on a category in a visual, it will interact with other visuals on the canvas. For example, in the visuals in Figure 1-64, I have clicked the "Italy" column in the "QUANTITY by WINE COUNTRY" chart, and this has highlighted the Italian wines in the "REVENUE by WINE" visual.

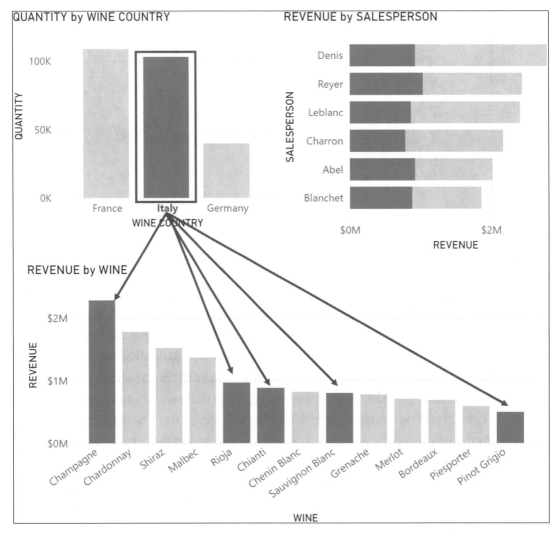

Figure 1-64. *The default interaction of a column chart*

However, if I now click "Denis" in the "REVENUE by SALESPERSON" bar chart, the interaction on the "REVENUE by WINE" chart will change to reflect "Denis's" sales and I have lost the "Italy" sales; see Figure 1-65.

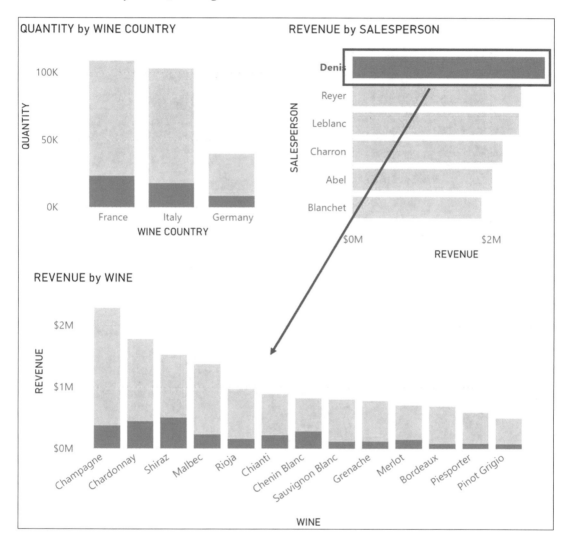

Figure 1-65. *Clicking another visual loses the previous interaction*

But what if I wanted to *retain* the Italy wine selection and show the sales for salesperson "Denis" *and* "Italy" wines in the "REVENUE by WINE" column chart? All I need to do is first click the "Italy" column, then hold down the **CTRL (or SHIFT)** key, and click the bar for "Denis" as shown in Figure 1-66.

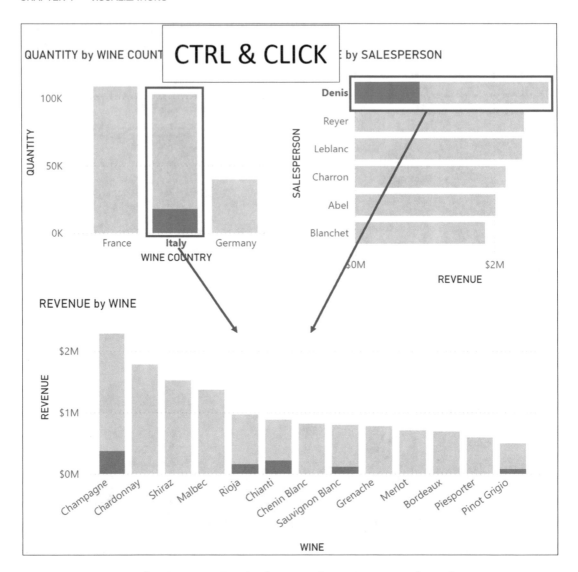

Figure 1-66. *Use the CTRL & CLICK key combination to multi-select*

I'm going to admit to you now that the reason I've included the answer to this question is that I'd been using Power BI for some years before someone showed me how I could do this. It was quite a revelation, I can tell you!

Q11. How Can I Help People Use My Visuals?

The answer is that you can add a Help tooltip to the Header icons. For example, you may have a visual that uses a drill down. Would people consuming your report know how to use the drill down icons in the visual header? If you doubt they will, then why not tell them? In the **Properties** section of the Format pane, find the **Header icons** card and the **Icons** subcard, turn on the **Help tooltip** icon, and in the **Tooltip text** box, type the text you want to show as help; see Figure 1-67.

Figure 1-67. *Turning on a Help tooltip*

The Help tooltip icon will sit in the header of the visual. People using the visual can hover on the icon to get the help they need. However, if you want to expand on the help text you provide, you may find the Help tooltip text area limited, for example, you can't format the text. Instead, you could create a Help tooltip page. To do this, on a new page (I named this page "Help Page") and in **Page information**, set the **Page type** to "Tooltip." This will generate a smaller page size on the canvas. Placing a text box on the Tooltip page means you can type information into this text box and format it accordingly. Then, in the Properties of the visual, in the Help tooltip card, set the Help tooltip **Type** to "Report page," and then find the page where you created the Tooltip in the **Page** dropdown. The steps in creating the Help tooltip Page are shown in Figure 1-68.

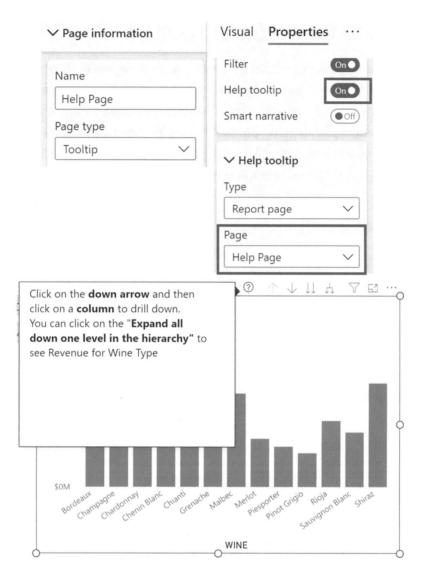

Figure 1-68. *Creating a Help tooltip page*

Using the Help tooltip or Help tooltip page, there's no need to keep consumers of your report in the dark. If in any doubt regarding their ability to use your report, let them know how they should interact with the visuals you provide.

Q12. Can One Visual Do the Job of Many Visuals?

One of the perennial problems when designing Power BI reports is how to fit all the visuals you require to tell the story of your data onto the report canvas; there just isn't enough room! Let's explore a scenario that illustrates this. You would like to show sales revenue by Salespeople, by Region, and by Product. For each of these categories, you would also like to show total, average, and maximum sales revenue. The conventional approach to show this data would result in nine visuals cluttering the canvas (Figure 1-69).

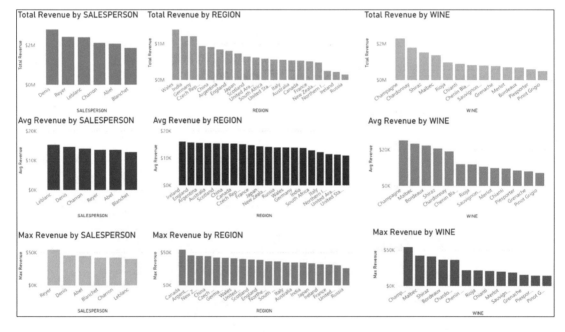

Figure 1-69. *The canvas can get quickly cluttered*

However, you could just have one visual and use slicers to toggle between the categories and metrics; see Figure 1-70.

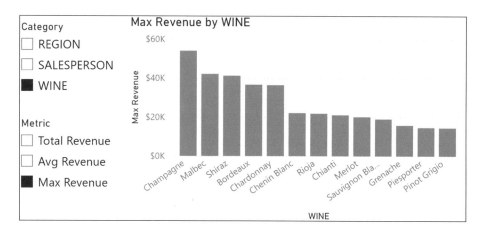

Figure 1-70. *One visual with slicers that toggle the data shown*

To achieve this happy state of affairs, you will need to create two "Fields" parameters.

Note   At the time of writing, the "Field Parameters" is a preview feature. To turn on Preview features, use the **File** tab, **Options and settings**, and **Options**. Then find the **Preview features** category under the GLOBAL settings.

The first of these will generate the slicer for the Categories. Select **Fields** from the **New parameter** button on the Modeling tab; see Figure 1-71.

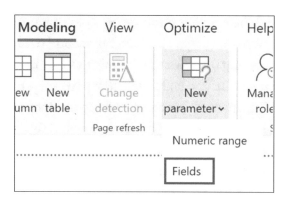

Figure 1-71. *The Fields option under the New parameter button*

In the **Name** box, you can rename the parameter, but we will just use the default "Parameter." Now, in "**Add and reorder fields**," you can drag and drop the fields that you want to comprise the parameter's slicer from the Fields list on the right. Note the option to "**Add a slicer to this page**" is turned on by default; see Figure 1-72.

What will your variable adjust?

Fields ⌄

Name

Parameter

Add and reorder fields

REGION ✕

SALESPERSON ✕

WINE ✕

☑ Add slicer to this page

Fields

🔍 Search

☐ SUPPLIER

☐ TYPE

☑ WINE

☐ WINE COUNTRY

☐ WINE ID

⌄ ▦ Regions

☑ REGION

☐ REGION ID

› ▦ Sales

⌄ ▦ SalesPeople

☐ FIRSTNAME

☑ SALESPERSON

☐ SALESPERSON ID

Figure 1-72. Drag and drop the fields you want in the parameter's slicer

Once you have clicked the **Create** button in the Fields parameter dialog box, the slicer will be placed on the report page. To create the second parameter for the Metrics slicer, repeat as above, selecting the measures you want to list in the slicer.

In the Data pane on the right, you will notice that two tables have been generated called "Parameter" and "Parameter 2," each containing a column also called "Parameter." I renamed these columns "Category" and "Metric," respectively.

Now create your visual that will use the parameters. In the example in Figure 1-70, we have used a clustered column chart. Into the X-axis bucket, we placed the Category column from the Parameter table, and we placed the Metric column from the Parameter 2 table into the Y-axis bucket; see Figure 1-73.

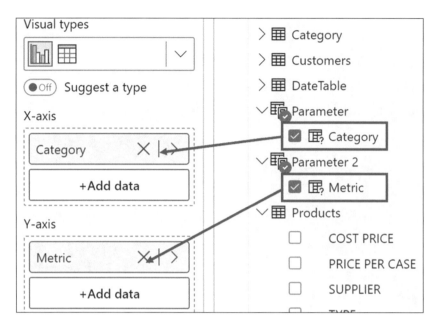

Figure 1-73. *Use the columns generated by the Parameters in the buckets of your visual*

Therefore, the answer to the question, "Can one visual do the job of many?" is a definite "yes." Using Fields parameters in this way means that you no longer have to clutter your canvas with a myriad of visuals. I'm also sure that you'll find numerous other uses for these parameters in your reports.

CHAPTER 2

Slicers and Filtering

One of the key ingredients of any Power BI report is the ability of consumers of the report to browse information by filtering the subsets of data that are important to them. However, designing reports where filters work in the way you want is not always straightforward. There are numerous ways that users can interact with a report, using slicers, using the Filters pane, or simply clicking on bars or columns in charts. In this chapter, we uncover some of the more obscure areas of filtering data that can improve the experience of the user when interacting with your reports.

Q13. Can a Slicer Highlight Data in a Visual?

Often, as Power BI professionals, we will categorically state that slicers only interact with visuals by *filtering* data, as opposed to clicking on charts that interact using a highlight. In Figure 2-1, for example, we have used a slicer to filter salesperson, "Charron," that interacts with the top column chart to filter Total Revenue for "Charron." However, if we use a bar chart to select the salesperson "Charron," this interacts with the column chart by highlighting Total Revenue for Charron.

© Alison Box 2023
A. Box, *A Power BI Compendium*, https://doi.org/10.1007/978-1-4842-9765-0_2

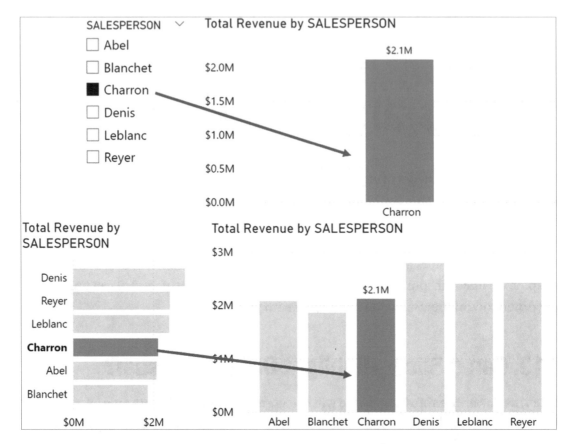

Figure 2-1. *Slicers filter data whereas chart visuals will highlight*

However, consider Figure 2-2, where we have selected "Charron" in the slicer and this has *highlighted* the column.

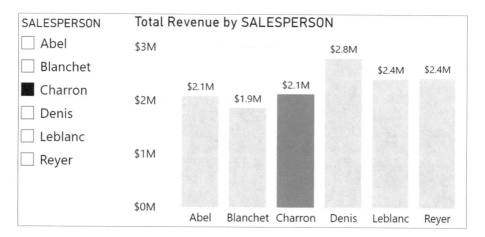

Figure 2-2. *A slicer highlighting a column*

If you want a slicer to highlight data in your visual and not filter it, then let's now explore how we achieved this visual interaction.

The slicer has to be populated with the SALESPERSON values. However, if we used this column from the Salespeople dimension, it would always *filter* the dimension, and we can't change this behavior. Therefore, we must create a duplicate table that contains just the values for this slicer, and the table will not be related to any other tables in the data model. To generate this table, you can use the **New Table** button on the **Modeling** tab and enter this DAX expression into the formula bar:

```
Salespeople Highlight = ALL ( Salespeople[SALESPERSON] )
```

This will create a table that contains the list of the salesperson's names; see Figure 2-3.

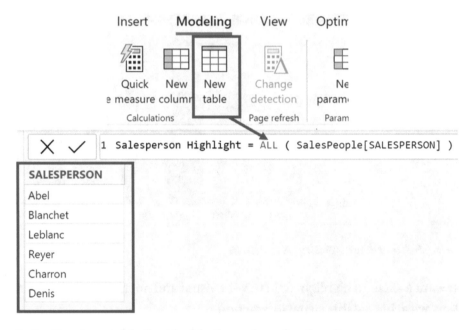

Figure 2-3. *Create a table that holds the values for the slicer*

Now we can use the SALESPERSON column from the Salesperson Highlight table in the slicer; see Figure 2-4.

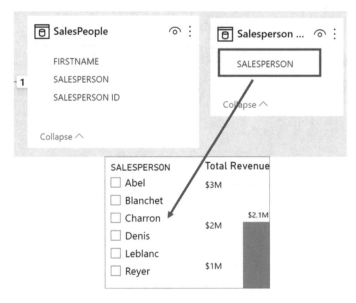

Figure 2-4. *Use the SALESPERSON column from the Salesperson Highlight table in the slicer*

The "highlight" colors (e.g., a pale blue and a darker blue) will be generated using conditional formatting on the column. Therefore, we must now create this simple DAX measure that will be used by the conditional formatting option:

```
Conditional Formatting Value =
VAR MySP =
  VALUES ( 'Salesperson Highlight'[SALESPERSON] )
RETURN
  IF ( SELECTEDVALUE ( SalesPeople[SALESPERSON] ) IN MySP, 1, 2 )
```

This measure uses the "MySP" variable to harvest the salespeople's names selected in the slicer. The measure will then return "1" if a column in the chart matches any values selected in the slicer; otherwise, it will return "2."

Now, in the **Format** pane, and using the **Columns** card, click the "*fx*" button beside the color option. In the Conditional formatting dialog box, in "**What field should we base this on?**" select the "Conditional Formatting Value" measure. Under **Rules**, choose a color to be used by a column if the value equals "1" (i.e., for the selected Salesperson). In our example, we chose the default blue. For the unselected values (i.e., equals "2"), we chose the pale blue color so that it mimicked the behavior of highlighting, but you could select any colors here; see Figure 2-5.

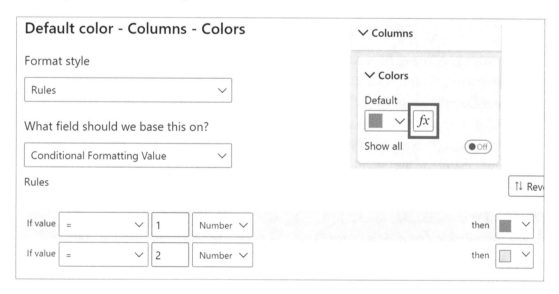

Figure 2-5. *Use conditional formatting to assign colors to the values returned by the "Conditional Formatting Value"*

However, you will have noticed that in this example the slicer has been filtering the same values that are in the column chart, that is, the Salesperson values. What if your objective is to use a slicer to filter a value from another dimension? For example, you may want to filter by year from the DateTable, and see each salesperson's values highlighted for the filtered year. Consider the column chart in Figure 2-6. You will see that we have selected the year 2022 in the slicer, and this has *highlighted* the data in the column chart.

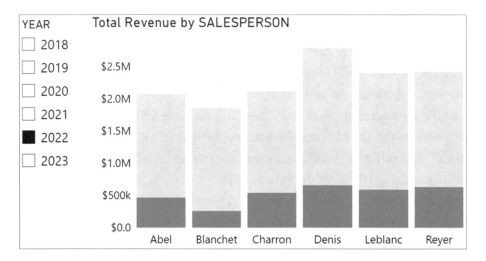

Figure 2-6. *A year is selected in the slicer and the data is highlighted*

Unfortunately, Power BI is not able to highlight data in this way using a slicer. We have to resort to a custom visual, and the visualization that will allow us to achieve our objective is Charticulator.

We've already met Charticulator when answering Q5 "How Can I Show Target Lines on Column Charts?" and just as in that example, we are going to use Charticulator's Data Axis to build the chart. To download Charticulator, refer to Q5.

Note   If you would like to learn more about Charticulator, you can read my book *Introducing Charticulator for Power BI*, published by Apress and available from Amazon.

Before creating the column chart, however, we must generate a table that will hold only the YEAR values from the DateTable. We do this in the same way we created the table holding the Salesperson's names and that was by using the **New Table** button on the **Modeling** tab. We can enter this DAX code:

```
Year Highlight = ALL ( DateTable[YEAR] )
```

Remember that this table will not be related to other tables in the data model. We can now put a slicer on the canvas and use the YEAR values from this table in the slicer. The next step is to create this DAX measure that will be used by Charticulator to highlight the data:

```
Conditional Formatting Year =
VAR MyYr =
    VALUES ( 'Year Highlight'[YEAR] )
RETURN
    CALCULATE([Total Revenue],DateTable[YEAR] in MyYr)
```

So that the value returned by this measure (i.e., the Total Revenue for the selected year) and the Total Revenue for all years is different, it's important that you now select a year in the slicer.

We can now select the Charticulator visual from the visualizations gallery and add the fields that will comprise the visual into the Data bucket. In our example, these are SALESPERSON, the Total Revenue measure, and the Conditional Formatting Year measure. Now click the **More options** button top right of the visual, and select **Edit** and then **Create a chart**. This will open Charticulator; see Figure 2-7.

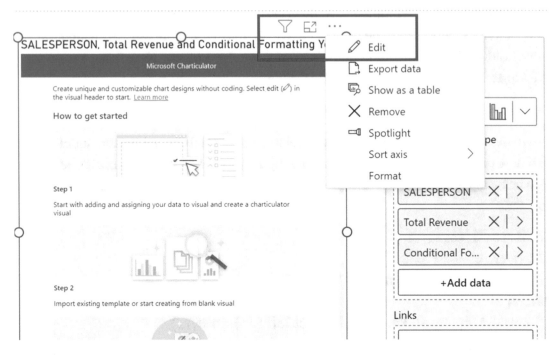

Figure 2-7. *Use the Edit button to open Charticulator*

From the menu bar in Charticulator, drag the **Data Axis** button into the **Glyph** pane. Then drag Total Revenue and Conditional Formatting Year (the values to be plotted) from the Fields pane onto the Data Axis in the **Glyph** pane; see Figure 2-8.

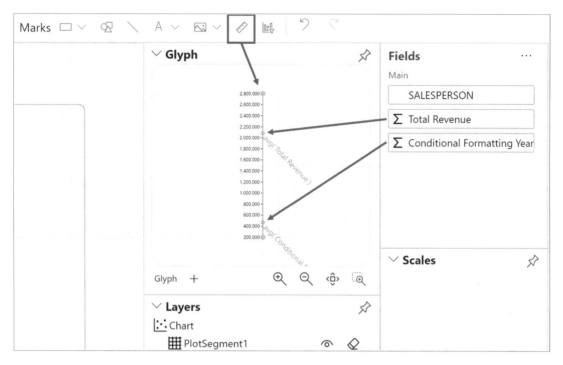

Figure 2-8. *Drag the Data Axis button into the Glyph pane, and drag the values onto the Data Axis*

We now need to create two columns, one on top of the other. One will be the highlighted column and the other the non-highlighted column. Click the **Marks** button, and draw a rectangle aligned to the "avg(Total Revenue)" mark on the Data Axis and within the gray dotted lines. This will be the non-highlighted column so it must be formatted with a pale color.

To format the column, in the **Layers** pane, select "Shape1," and then in the **Attributes** pane and in the **Fill** option, click the fill color to the right of the box and select a pale color; see Figure 2-9.

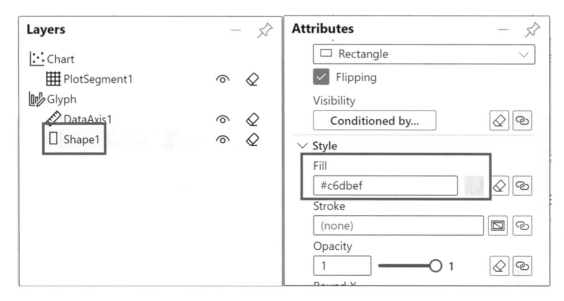

Figure 2-9. *Format the rectangle with a pale Fill color*

Now draw another rectangle in the Glyph pane, aligned to "avg(Conditional Formatting Year)." If you are happy with the default blue, you don't need to change the fill color. The **Glyph** pane should look similar to Figure 2-10.

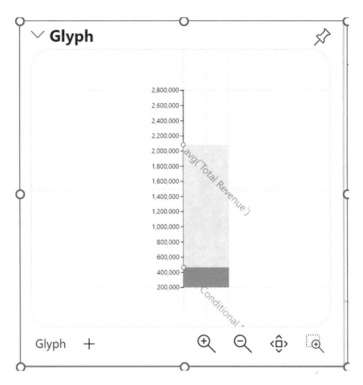

Figure 2-10. *The Glyph pane with both rectangles created*

To show the salespeople's names on the X-axis, drag the SALESPERSON field onto the X-axis on Charticulator's chart canvas (Figure 2-11). You may also need to drag on the gray guide lines to pull the X- and Y-axes onto the canvas, shown by red ovals in Figure 2-11.

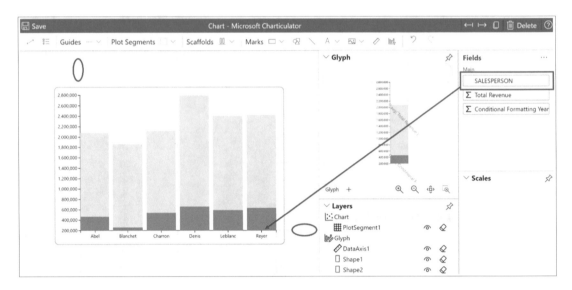

Figure 2-11. *Drag the SALESPERSON onto the X-axis*

Now click the Save button top left of Charticulator, and click Back to report (also top left) to return to Power BI.

Why not now have a browse around Charticulator's other options? Select a chart element in the **Layers** pane, and the **Attributes** pane will be listed all the formatting options for that element. For example, you may want to change the starting value on the Data Axis to zero.

So the answer to "Can a Slicer Highlight Data?" is yes it can, but I think you'll agree that it's not as straightforward as we might have hoped.

Q14. How Can I Stop Values in Slicers "Sticking"?

This can be a problem if you have cascading slicers in a "parent/child" scenario. When you select an item in the "parent" slicer, it will filter "children" of that "parent" in the "child" slicer. What you don't want, however, is to select items in the "child" slicer and the "parent" slicer responds to this filter. Therefore, you edit the interactions of the "child" slicer accordingly. Unfortunately, what will now happen is when you select a "child" item, it will "stick" on a re-selection in the "parent" slicer when it is now no longer a "child" of that "parent."

As an example of this behavior, we can use data from the Products table where we have a WINE column and a TYPE column. Each wine has a type, either "Red" or "White"; see Figure 2-12.

WINE ID	WINE	SUPPLIER	TYPE	WINE
1	Bordeaux	Laithwaites	Red	France
2	Champagne	Laithwaites	White	Italy
3	Chardonnay	Alliance	White	Germa
4	Malbec	Laithwaites	Red	Germa
5	Grenache	Redsky	Red	France
13	Shiraz	Alliance	Red	France
6	Piesporter	Redsky	White	France
11	Rioja	Majestic	Red	Italy
7	Chianti	Redsky	Red	Italy
12	Chenin Blanc	Alliance	White	France
8	Pinot Grigio	Majestic	White	Italy
9	Merlot	Majestic	Red	France
10	Sauvignon Blanc	Majestic	White	Italy
14	Lambrusco	Alliance	White	Italy

Figure 2-12. *Each wine is either "Red" or "White"*

Therefore, when we put slicers on the canvas, a slicer using the TYPE column is our "parent," and a slicer using the WINE column is our "child." We now must edit the interactions of the WINE slicer so it won't filter the TYPE slicer.

Now when we select "White" in the TYPE slicer, the WINE slicer correctly shows only White wines. But if we now select "Piesporter" in the WINE slicer and then select "Red" in the TYPE slicer, "Piesporter" is still listed even though it is not a red wine (see Figure 2-13).

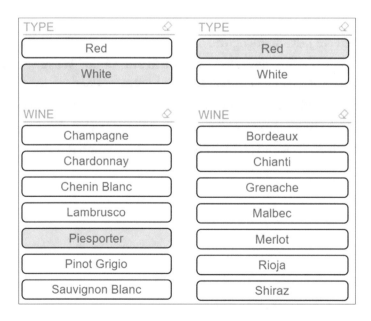

Figure 2-13. *Selecting "Red" still shows "Piesporter"*

To avoid this issue, use a Chiclet slicer. This is a free custom visual available to download from the Microsoft App Store, and it won't exhibit this behavior; see Figure 2-14.

Figure 2-14. *The Chiclet slicer won't show items that are not filtered*

When using the Chiclet slicer, ensure that on the **General** formatting card, the option for "Multiple Selection" has been turned off.

Q15. How Do I Generate Partial Matches in a Slicer?

One answer to this question is to turn on the Search box on a slicer and type your partial match into this. Click the **More options** button of the slicer and select **Search** from the options. This places a Search box at the top of a slicer. In Figure 2-15, I have typed "c" into the Search box, and the slicer has found any value that contains "c."

Figure 2-15. *You can use the Search box to find partial matches*

The problem is that we would now have to select some or all of these values in this slicer to filter only those values we required. Consider Figure 2-16 which is a preferable solution. Here we have generated a slicer that lists letters to provide a match for either values that begin with this letter or contain the letter, and the wine names are filtered accordingly.

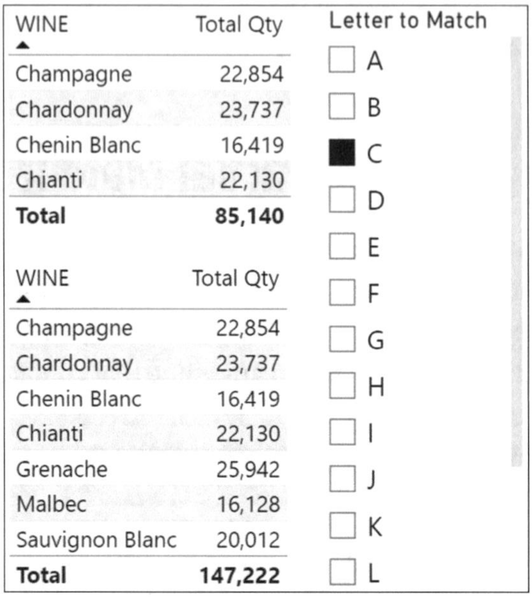

WINE	Total Qty
Champagne	22,854
Chardonnay	23,737
Chenin Blanc	16,419
Chianti	22,130
Total	**85,140**

WINE	Total Qty
Champagne	22,854
Chardonnay	23,737
Chenin Blanc	16,419
Chianti	22,130
Grenache	25,942
Malbec	16,128
Sauvignon Blanc	20,012
Total	**147,222**

Letter to Match

☐ A
☐ B
■ C
☐ D
☐ E
☐ F
☐ G
☐ H
☐ I
☐ J
☐ K
☐ L

Figure 2-16. *Using a slicer that lists letters will filter data accordingly*

To create the slicer, you can use the **Enter Data** button on the **Home** tab and create a table that has a single column that lists the letters alphabetically. I named my table "Slicer" and the column containing the letters, "Letter to Match."

Now to create the DAX measure that will be used to filter the data:

```
Match Letter =
FIND ( SELECTEDVALUE ( Slicer[Letter to Match] ),
        SELECTEDVALUE ( Products[WINE] ), ,0 )
```

We've used the FIND function here that will match the letter selected in the slicer to the letters within the wine names and return that letter's position within the text string. The FIND function is case sensitive so, for example, selecting "C" in the slicer will return 1 for "Champagne" but will return 0 for "Malbec." If searching for "C" or "c" is important to you, you can use the SEARCH function which is case insensitive instead:

```
Match Letter 2 =
SEARCH ( SELECTEDVALUE ( Slicer[Letter to Match] ),
        SELECTEDVALUE ( Products[WINE] ),1,0 )
```

Now we've created the measures we need, they must be placed in a visual level filter in the **Filters** pane. If you are using the FIND function, you must filter values that equal 1. If using the SEARCH function, you filter values greater than 0, see Figure 2-17.

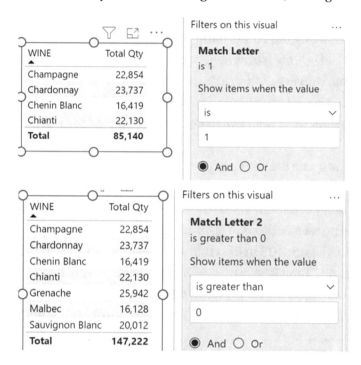

Figure 2-17. *Use visual level filters to filter the values*

The key factor to the solution to this question is using DAX measures to return values based on the position of letters within the names of the wines. We can then simply filter the desired position of that letter.

Q16. How Do You Filter Using a DAX Measure?

A question that we are often asked is this: Can you put a measure into a slicer so you can filter by the values it returns? The answer of course is no; only columns from tables can be populated into slicers.

Note   In Q12 "How Can One Visual Do the Job of Many Visuals," we explored using Fields Parameters in slicers, but these allow you to filter using different measures, not values returned by measures.

The reason why this question so often arises is that most people see the slicer as the most important filtering tool on the report.

However, if you want to filter data using values returned by measures, this is where the visual level filter using the Filters pane comes into its own. One of the defining differences between using the Filters pane and using slicers is that in the Filters pane buckets, you can put DAX measures to generate visual level filtering and also page or report level filtering. However, when using measures to filter data, you will mostly use a visual level filter to constrain the filter to specific visuals. What's more, the measure on which you want to filter does not have to be placed in any of the buckets that comprise the visual.

Let's now look at some examples of using measures in the visual level filter bucket of the Filters pane.

If you want to analyze data for which there are no values to show, for example, products for which there are no sales, using a measure in the visual level filter makes this analysis easy. For example, in Figure 2-18 in the table visual we are using the WINE column from a dimension table and "Total Revenue" is a DAX measure. We are looking at the data for January 2022 and have used the "Show items with no data" option on the WINE column of the table visual to see the wines that have no Total Revenue for this month.

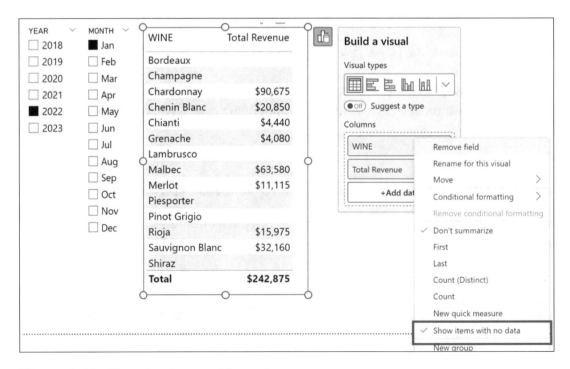

Figure 2-18. Showing items with no data

If we now only want to show items with no data in the visual, we can use the visual level filter in the Filters pane to filter values that are blank. We can then remove the "Total Revenue" measure from the Table visual because there are no values to show; see Figure 2-19.

Figure 2-19. *A visual level filter can show only items with no data*

Another example of filtering using measures is where you may want to filter text values returned by a DAX measure. For example, you may have a measure, for example, "ABC Customers," that categorizes your customers into "A," "B," or "C" based on ABC analysis (see Q51, "How Do I Calculate Pareto or ABC Analysis"). You can browse your products by classification on an ad hoc basis by using a visual level filter to filter the customers accordingly; see Figure 2-20.

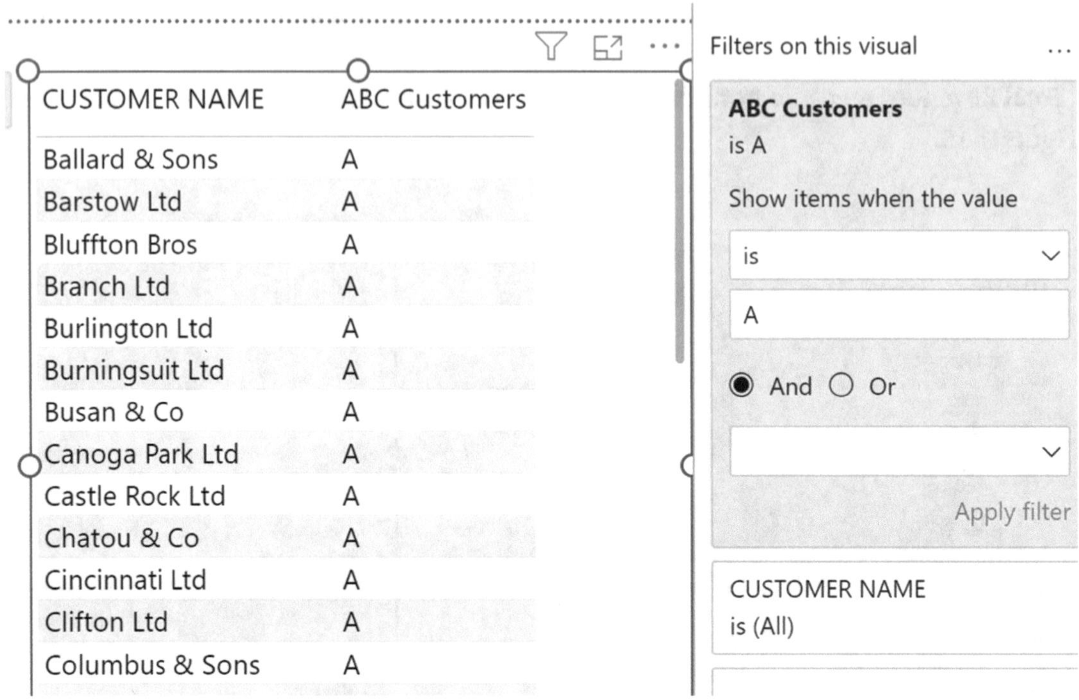

CUSTOMER NAME	ABC Customers
Ballard & Sons	A
Barstow Ltd	A
Bluffton Bros	A
Branch Ltd	A
Burlington Ltd	A
Burningsuit Ltd	A
Busan & Co	A
Canoga Park Ltd	A
Castle Rock Ltd	A
Chatou & Co	A
Cincinnati Ltd	A
Clifton Ltd	A
Columbus & Sons	A

Filters on this visual

ABC Customers
is A

Show items when the value

is

A

⦿ And ◯ Or

Apply filter

CUSTOMER NAME
is (All)

Figure 2-20. *Filtering text values returned by a measure*

So far, we've explored two examples of filtering values returned by DAX measures, filtering items with no data and filtering text values, and undoubtedly both can be beneficial to your data analysis. However, the most obvious reason to filter a measure is to filter numeric values. If this is your goal, all is not plain sailing. For example, consider Figure 2-21. In the matrix visual, we have the "YEAR" and "QTR" fields sitting in the "rows" of the matrix. We are using the "Total Qty" measure in the visual level filter card and filtering values that are greater than 15,000. This works fine all the while the values being filtered sit in the "rows" of the matrix.

Figure 2-21. *A visual level filter filtering values in the rows greater than 15,000*

However, if we move the "QTR" field onto the "columns" of the matrix, the filter appears to no longer work; see Figure 2-22.

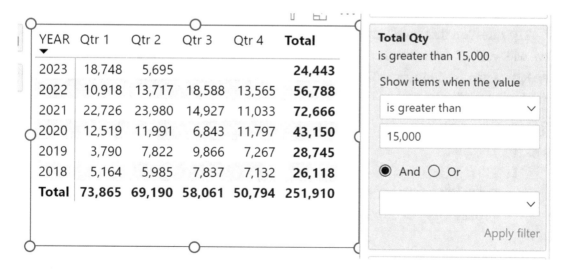

Figure 2-22. *A visual level filter filtering values in the columns greater than 15,000*

In fact, the filter is working; it's filtering the Total values sitting in the YEAR rows of the matrix. Therefore, if we change the filter criterion to "greater than 30,000," we can see the filter is correct (Figure 2-23).

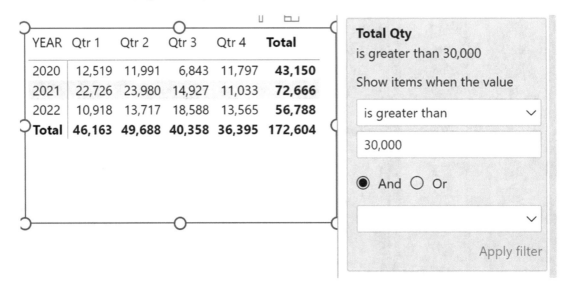

Figure 2-23. *A visual level filter filtering values in the rows greater than 30,000*

What we must understand from this scenario is that the visual level filter is only applied to values sitting in the "rows" of a Matrix visual, not to values sitting in the "columns."

The question that now remains is how we can filter quantities greater than 15,000 that will work when we use the "columns" bucket of a matrix visual. For this, we can't use the Filters pane but must use DAX instead. This would be the measure that would filter our data accordingly:

```
Filtered Qty =
SUMX (
    FILTER (
        SUMMARIZE ( DateTable, DateTable[YEAR], DateTable[QTR] ),
        [Total Qty] > 15000
    ),
    [Total Qty]
)
```

If we put this measure into the "Values" bucket of the Matrix, we now have the correct filtered values; see Figure 2-24.

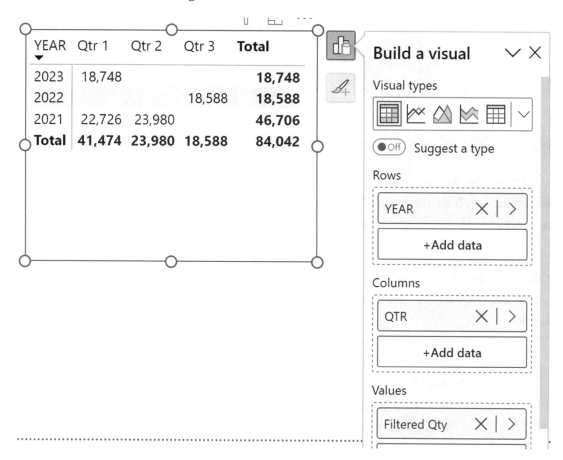

Figure 2-24. *Using a DAX measure to generate a filter*

But of course, we are now limited to viewing quantities greater than 15,000, whereas we might want to filter different quantities. To remedy this, you could create a parameter table, using the **Enter data** button on the **Table tools** tab that holds various filtering criteria, as shown in Figure 2-25.

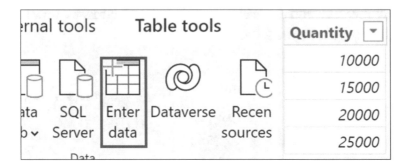

Figure 2-25. *Create a parameter table with the values you want to filter by*

We named our parameter table "Select Qty." Parameter tables are not related to any other tables in the data model. You could then populate a slicer with the "Quantity" column from this table and simply edit the "Total Qty" measure as follows:

```
Filtered Qty =
VAR qtyvalue =
    SELECTEDVALUE ( 'Select Qty'[Quantity] )
RETURN
    SUMX (
        FILTER (
            SUMMARIZE ( DateTable, DateTable[Year],
                DateTable[Qtr] ),
            [Total Qty] > qtyvalue
        ),
        [Total Qty]
    )
```

Now we can browse the data, selecting different criteria by which to filter the Total Qty values, Figure 2-26.

YEAR	Qtr 1	Qtr 2	Qtr 3	Qtr 4	Total
2018	5,164	5,985	7,837	7,132	**26,118**
2019		7,822	9,866	7,267	**24,955**
2020	12,519	11,991	6,843	11,797	**43,150**
2021	22,726	23,980	14,927	11,033	**72,666**
2022	10,918	13,717	18,588	13,565	**56,788**
2023	18,748	5,695			**24,443**
Total	**70,075**	**69,190**	**58,061**	**50,794**	**248,120**

Quantity
- ■ 5,000
- ☐ 10,000
- ☐ 15,000
- ☐ 20,000

Figure 2-26. *Use a slicer to filter values in the Matrix*

So the answer to how do you filter data using a DAX measure is that, under most circumstances, you would use a visual level filter in the Filters pane. However, we have also seen that it is possible to use slicers populated with a column from a parameter table.

Q17. How Do I Cross Filter Slicers?

In a typical data model that is based on a star schema where dimension tables are directly related to the fact table, using a slicer to filter values in one dimension will not cross-filter values in another. Filters only flow from the one side of a relationship to the many, indicated by the arrow in Model view; see Figure 2-27.

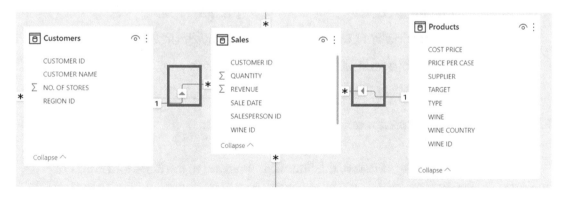

Figure 2-27. *Filters only flow in the direction of the arrows*

For example, in Figure 2-28, you can see we have two slicers, one using the CUSTOMER NAME column from the Customers dimension and one using the WINE column from the Products dimension. If we select a value in the CUSTOMER NAME slicer, that is, "Ballard & Sons," the WINE slicer won't change to reflect the wines that "Ballard & Sons" has bought. We always see all the wines regardless of selections made in the CUSTOMER NAME slicer.

CUSTOMER NAME	WINE
☐ Acme & Sons	☐ Bordeaux
■ Ballard & Sons	☐ Champagne
☐ Barstow Ltd	☐ Chardonnay
☐ Beaverton & Co	☐ Chenin Blanc
☐ Black Ltd	☐ Chianti
☐ Black River & Co	☐ Grenache
☐ Bloxon Bros.	☐ Lambrusco
☐ Bluffton Bros	☐ Malbec
☐ Branch Ltd	☐ Merlot
☐ Brooklyn & Co	☐ Piesporter
☐ Brooklyn Ltd	☐ Pinot Grigio
☐ Brown & Co	☐ Rioja
☐ Burlington Ltd	☐ Sauvignon Blanc
☐ Burningsuit Ltd	☐ Shiraz
☐ Busan & Co	

Figure 2-28. *Filtering customers does not filter wines*

If the Customers table is filtered, the filter is propagated to the Sales table but not filtered onward to the Products table because filters don't flow from the many side of the relationship to the one side. However, we can force the Products table to cross-filter accordingly. We can do this by placing a visual filter on the WINE slicer using a *measure*, such as "Total Revenue" and filtering only Wines that have a "Total Revenue" value. In fact, we can use any measure that performs a calculation on the Fact table and then set this filter to "Show items when the value is not blank" as shown in the visual filter in Figure 2-29.

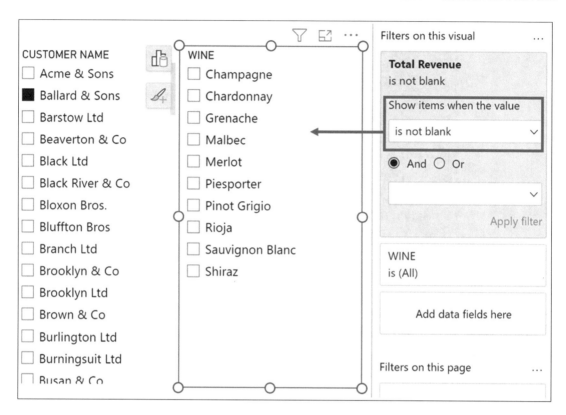

Figure 2-29. *To cross-filter slicers, use a visual filter populated with a measure and set to "is not blank"*

This particular visual level filter works in much the same way as the "Show items with no data" option that we use in a matrix or table visual where we would apply this option to the WINE value in the visual. It's just that this filter is reversing this option to find items where there *is* data. The takeaway here is how important the star schema is when designing Power BI data models. By generating DAX measures, using values from clearly defined dimensions and fact tables enables us to filter by these measures and eliminate items either where there is data or where there is no data.

Q18. How Can I Create an "And" Slicer?

The problem with slicers is that they filter using "OR" criteria. Consider Figure 2-30. Here we have used a slicer to filter customers who have bought "Bordeaux," "Champagne," or "Chardonnay." It will filter customers who have bought any or all of these products. We must find out which customers only bought *all* products and find the value for the "Total Qty" measure just for these customers.

CUSTOMER NAME	Total Qty		WINE
Ballard & Sons	935		■ Bordeaux
Barstow Ltd	957		■ Champagne
Black Ltd	315		■ Chardonnay
Black River & Co	200		☐ Chenin Blanc
Bluffton Bros	588		☐ Chianti
Branch Ltd	1,810		☐ Grenache
Brooklyn Ltd	343		☐ Lambrusco
Brown & Co	993		
Burlington Ltd	371		
Burningsuit Ltd	925		
Busan & Co	854		
Canoga Park Ltd	442		

Figure 2-30. *The slicer filters customers who have bought "Bordeaux,"
"Champagne," or "Chardonnay," or any of these*

To do this we must generate this measure:

```
Filtering AND =
VAR myproducts =
    DISTINCTCOUNT( Sales[WINE ID] )
VAR noOfwines =
    COUNTROWS ( Products ) VAR allwines = COUNTROWS ( ALL ( Products ) )
RETURN
    IF (myproducts = noOfwines || allwines = noOfwines, noOfwines )
```

The "myproducts" variable counts the number of different products that each
customer has purchased within the products filtered in the slicer. The "noOfwines"
variable counts the number of products selected in the slicer. If these values match, the
measure returns the number of products selected in the slicer. If they don't match it
returns blank. The "allwines" variable checks for no selections in the slicer.

We must now use this measure in a visual level filter to filter values that are not blank
and so arrive at a list of correct customers. In Figure 2-31, we can see that our customers
purchased **10,127** cases of "Bordeaux," "Champagne," *and* "Chardonnay."

Figure 2-31. *The "Filtering AND" measure returns a count of the wines selected in the slicer but only for the customers who have bought those wines*

Therefore, if your requirement is to filter items that match *all* selections in the slicer and not just any of them, you can use this simple piece of code.

Q19. Is It Possible to See the Source Records When Browsing a Visual?

Yes, it is, so be vigilant. Under certain circumstances, you're given the option to "Show data point as a table" which exposes the source records that comprise the data point. If you're browsing a column chart, for example, that uses an implicit measure in the Y-axis bucket, you can right-click a column, and this option is available on the shortcut menu; see Figure 2-32.

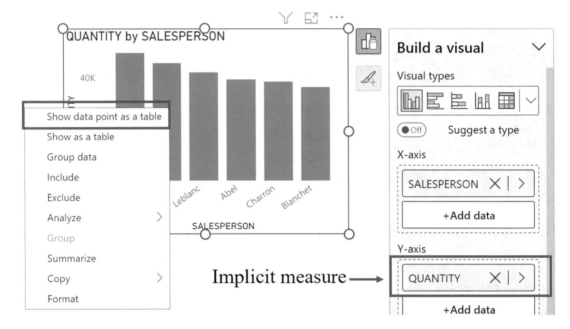

Figure 2-32. *If you are using an implicit measure, you have the option to "Show data point as a table" on the shortcut menu*

Implicit measures are identified by the sigma symbol (Σ) that sits beside them in the Data pane. For example, in Figure 2-32, we have used the QUANTITY implicit measure from the Sales table. The data point table is shown in Figure 2-33.

SALESPERSON	QUANTITY	SALE DATE	CUSTOMER NAME
Leblanc	81	16 November 2020	Bluffton Bros
Leblanc	98	11 March 2018	Erlangen & Co
Leblanc	98	21 December 2019	Erlangen & Co
Leblanc	99	17 March 2018	Kennebunkport & Co
Leblanc	102	21 November 2020	Charlottesville & Co
Leblanc	106	10 January 2018	Melbourne Ltd
Leblanc	106	11 August 2020	Melbourne Ltd
Leblanc	106	23 August 2020	Melbourne Ltd
Leblanc	106	14 July 2021	Colombes & Co
Leblanc	112	19 September 2018	Saint Germain en Laye & Co
Leblanc	113	20 September 2021	Charlottesville & Co

Figure 2-33. *A data point table*

However, you will notice that we're only provided with a small selection of columns, and unfortunately, there is no way to customize this table. If you want to see additional columns in the data point table, you must edit the visual to include those columns. For example, in Figure 2-34, we have included the WINE column from the Products dimension.

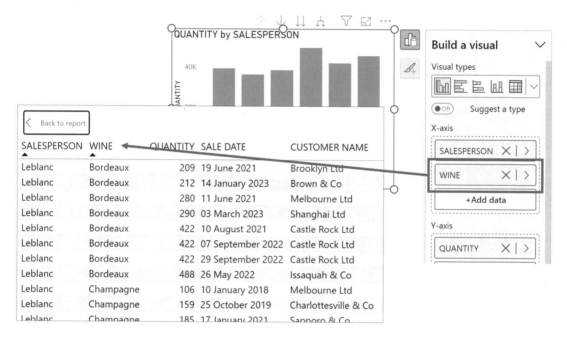

Figure 2-34. *Add additional columns to your visual to be included in the data point table*

Clearly, being able to view the underlying data is not always desirable. If you create your own measures using DAX, then the data point table is not available. This is a compelling reason to use calculated measures in your reports and not the implicit variety.

CHAPTER 3

The Matrix

About seven years ago, I published a blog post entitled "7 Secrets of the Matrix Visual." Over the years, this blog has far outstripped any of my other posts with regard to the traffic it has attracted.

To read this blog post, visit `www.burningsuit.co.uk/blog/7-secrets-matrix-visual`.

I have often wondered why this has been the case and have concluded that Power BI's matrix is perhaps the most unintuitive and difficult visualization to work with. Plowing through the myriad of options and settings that must be edited to effect a simple change in design can leave you feeling frustrated, if not exhausted. In this chapter, we answer specific questions that throw light on some of the darker corners of the Power BI matrix to alleviate these frustrations and enable you to design a matrix that truly complies with the analysis of your data.

Q20. How Can I Control Which Subtotals Show?

This is such a great question because it should be a straightforward task! The matrix visual is Power BI's equivalent of Excel's PivotTable, and therefore, you might think that controlling the display of grand totals and subtotals in the matrix should be as simple as in the PivotTable where you can use dropdown menus or specific Field settings (Figure 3-1).

© Alison Box 2023
A. Box, *A Power BI Compendium*, https://doi.org/10.1007/978-1-4842-9765-0_3

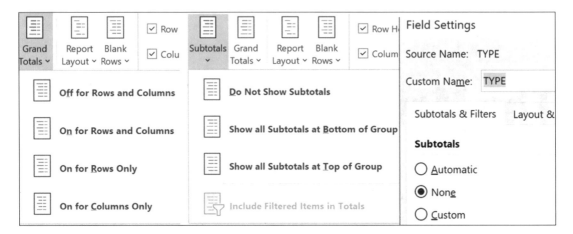

Figure 3-1. *Excel PivotTable options for grand totals and subtotals are easy to apply*

However, in the Power BI matrix visual, things are not quite as simple. Let's take a look at some of the difficulties you may encounter when attempting to control calculations at different levels in the matrix and how we can resolve them. First, we will construct a matrix visual using our sales data as an example. The matrix will comprise the WINE, TYPE, and SALESPEOPLE fields in the Rows bucket and the YEAR field in the Columns bucket. The Total Revenue measure will be put into Values; see Figure 3-2. You must use the **Expand all down one level in the hierarchy** button in the visual header to see the subtotals.

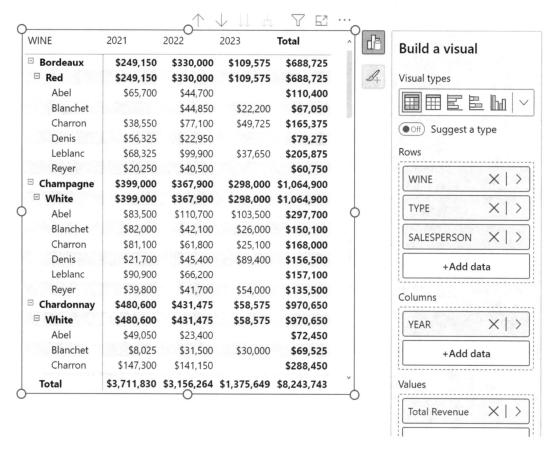

WINE	2021	2022	2023	Total
⊟ **Bordeaux**	**$249,150**	**$330,000**	**$109,575**	**$688,725**
⊟ **Red**	**$249,150**	**$330,000**	**$109,575**	**$688,725**
Abel	$65,700	$44,700		**$110,400**
Blanchet		$44,850	$22,200	**$67,050**
Charron	$38,550	$77,100	$49,725	**$165,375**
Denis	$56,325	$22,950		**$79,275**
Leblanc	$68,325	$99,900	$37,650	**$205,875**
Reyer	$20,250	$40,500		**$60,750**
⊟ **Champagne**	**$399,000**	**$367,900**	**$298,000**	**$1,064,900**
⊟ **White**	**$399,000**	**$367,900**	**$298,000**	**$1,064,900**
Abel	$83,500	$110,700	$103,500	**$297,700**
Blanchet	$82,000	$42,100	$26,000	**$150,100**
Charron	$81,100	$61,800	$25,100	**$168,000**
Denis	$21,700	$45,400	$89,400	**$156,500**
Leblanc	$90,900	$66,200		**$157,100**
Reyer	$39,800	$41,700	$54,000	**$135,500**
⊟ **Chardonnay**	**$480,600**	**$431,475**	**$58,575**	**$970,650**
⊟ **White**	**$480,600**	**$431,475**	**$58,575**	**$970,650**
Abel	$49,050	$23,400		**$72,450**
Blanchet	$8,025	$31,500	$30,000	**$69,525**
Charron	$147,300	$141,150		**$288,450**
Total	**$3,711,830**	**$3,156,264**	**$1,375,649**	**$8,243,743**

Build a visual

Visual types

● Off Suggest a type

Rows

WINE   ✕ | 〉

TYPE   ✕ | 〉

SALESPERSON   ✕ | 〉

+Add data

Columns

YEAR   ✕ | 〉

+Add data

Values

Total Revenue   ✕ | 〉

Figure 3-2. *The Power BI matrix comprising Rows, Columns, and Values*

The important factor in this matrix is that each WINE value has a one-to-one correspondence to its TYPE. For example, the wine, "Bordeaux," will be "Red," and "Champagne" will be "White." This is why there is a duplication of subtotals. Clearly, the pressing requirement here is how to remove one of these subtotals.

The first problem that we encounter is that if we turn off the **Column** and **Row subtotals** by using the respective cards in the **Format** pane, this also removes the grand totals (Figure 3-3).

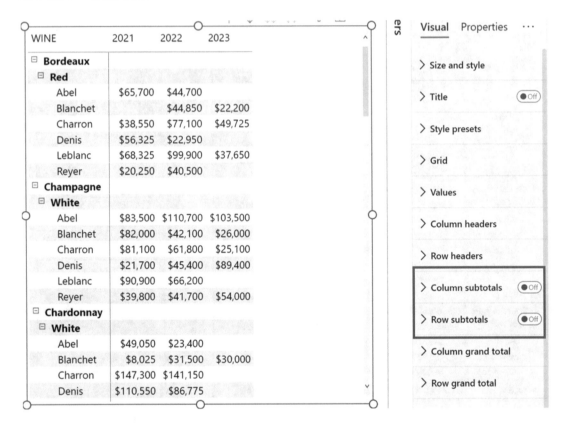

Figure 3-3. *Turning off subtotals also removes grand totals*

It may be that you don't want to show the column grand totals, but you do want to show the row subtotals. To retain row subtotals but not column grand totals, follow these steps:

1. Turn on the **Column subtotals.**

2. Expand the **Columns subtotals** card.

3. In the **Apply Settings to** subcard, turn on **Per column level**.

4. In the **Column level** dropdown, select "YEAR."

5. In the **Columns** subcard, turn off **Show subtotal**.

Yes, it really is this complex! See Figure 3-4. To remove the row grand total, you would follow the same process on the **Row subtotals** card, selecting the WINE subtotal as the row level to turn off.

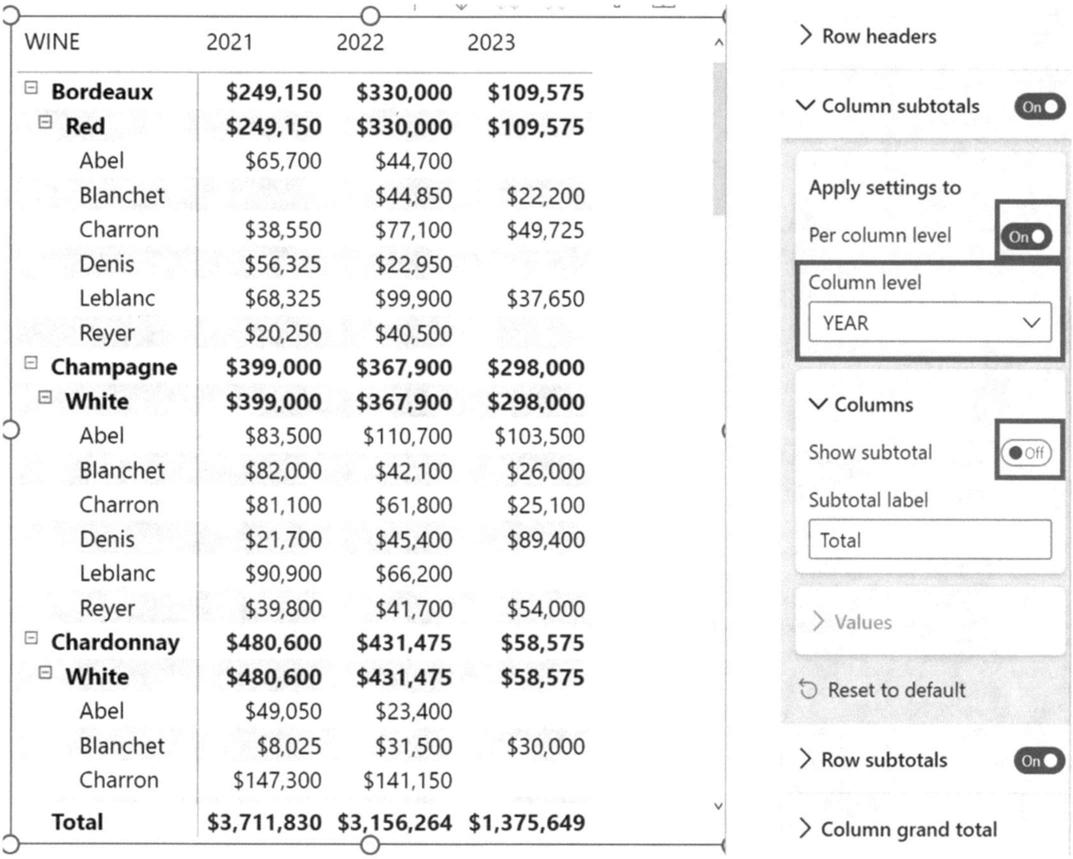

Figure 3-4. *Turning off the grand totals but retaining subtotals*

However, we still have the problem of the duplication of subtotals, for the TYPE and the SALESPERSON columns. Clearly, one of these must go! To turn off a specific subtotal, take these steps as shown in Figure 3-5:

1. Under the **Row subtotals** card, in the **Apply settings to** subcard, turn on **Per row level**.

2. In the **Row level** dropdown, select, for example, SALESPERSON.

3. In the **Rows** subcard, turn off these subtotals.

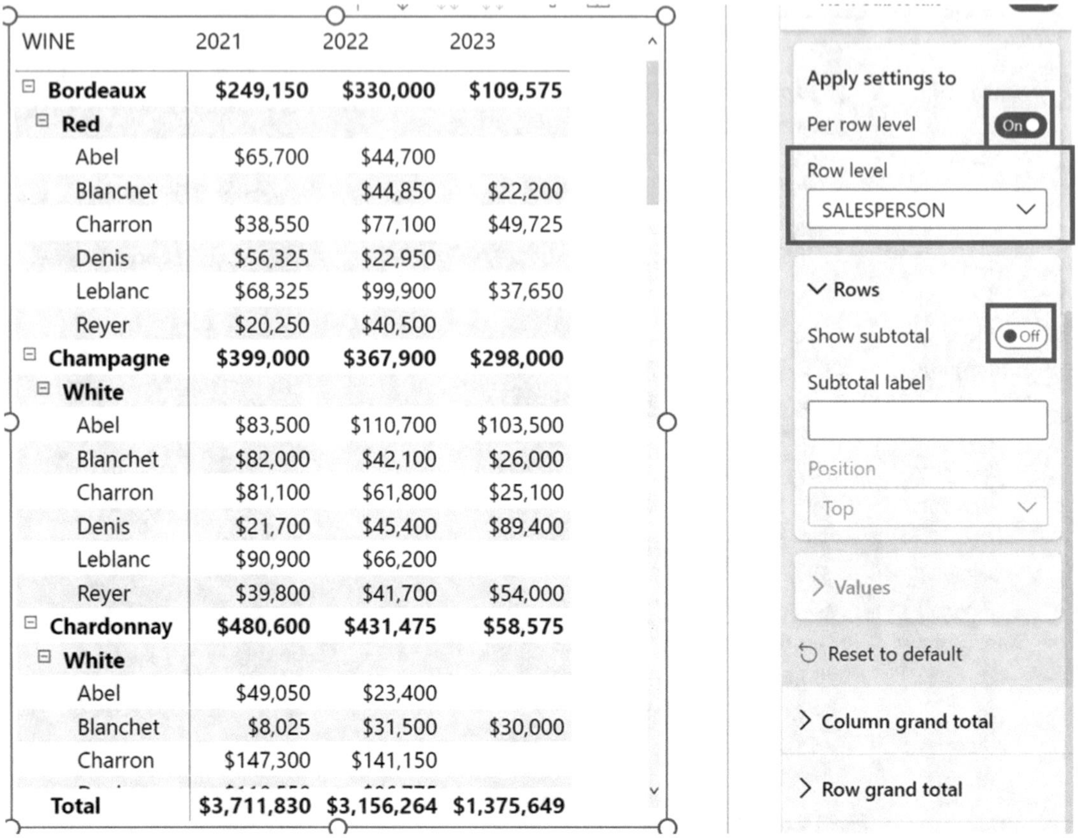

WINE	2021	2022	2023
⊟ **Bordeaux**	**$249,150**	**$330,000**	**$109,575**
⊟ **Red**			
Abel	$65,700	$44,700	
Blanchet		$44,850	$22,200
Charron	$38,550	$77,100	$49,725
Denis	$56,325	$22,950	
Leblanc	$68,325	$99,900	$37,650
Reyer	$20,250	$40,500	
⊟ **Champagne**	**$399,000**	**$367,900**	**$298,000**
⊟ **White**			
Abel	$83,500	$110,700	$103,500
Blanchet	$82,000	$42,100	$26,000
Charron	$81,100	$61,800	$25,100
Denis	$21,700	$45,400	$89,400
Leblanc	$90,900	$66,200	
Reyer	$39,800	$41,700	$54,000
⊟ **Chardonnay**	**$480,600**	**$431,475**	**$58,575**
⊟ **White**			
Abel	$49,050	$23,400	
Blanchet	$8,025	$31,500	$30,000
Charron	$147,300	$141,150	
Total	**$3,711,830**	**$3,156,264**	**$1,375,649**

Apply settings to

Per row level   On ●

Row level

SALESPERSON   ⌄

⌄ Rows

Show subtotal   ● Off

Subtotal label

Position

Top   ⌄

❯ Values

↻ Reset to default

❯ Column grand total

❯ Row grand total

Figure 3-5. *Turning off the subtotal for the SALESPERSON column*

However, it looks odd that the matrix now has an empty row where the subtotal used to be displayed. To resolve this problem, in the **Row headers** card, we can turn off **Stepped layout** under the **Options** subcard; see Figure 3-6.

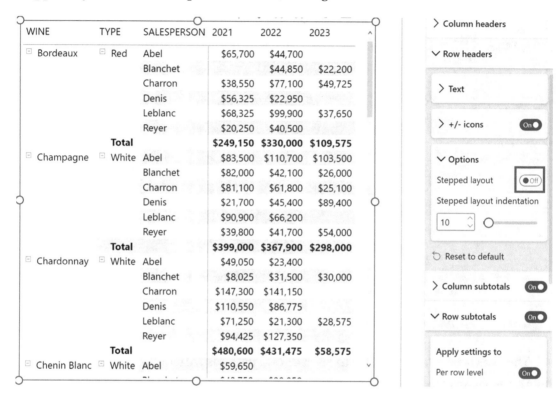

Figure 3-6. *Turning off the stepped layout*

This also moves the subtotal to the bottom of the group, although you can change this by using the **Position** option under the **Rows** subcard as shown in Figure 3-7.

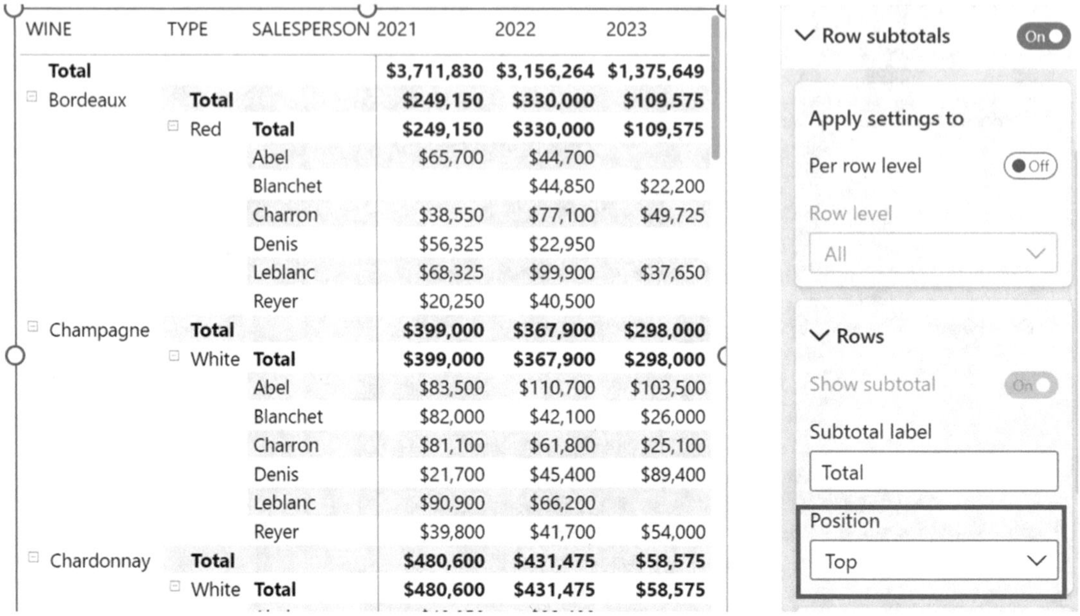

WINE	TYPE	SALESPERSON	2021	2022	2023
Total			**$3,711,830**	**$3,156,264**	**$1,375,649**
Bordeaux	**Total**		**$249,150**	**$330,000**	**$109,575**
	Red	**Total**	**$249,150**	**$330,000**	**$109,575**
		Abel	$65,700	$44,700	
		Blanchet		$44,850	$22,200
		Charron	$38,550	$77,100	$49,725
		Denis	$56,325	$22,950	
		Leblanc	$68,325	$99,900	$37,650
		Reyer	$20,250	$40,500	
Champagne	**Total**		**$399,000**	**$367,900**	**$298,000**
	White	**Total**	**$399,000**	**$367,900**	**$298,000**
		Abel	$83,500	$110,700	$103,500
		Blanchet	$82,000	$42,100	$26,000
		Charron	$81,100	$61,800	$25,100
		Denis	$21,700	$45,400	$89,400
		Leblanc	$90,900	$66,200	
		Reyer	$39,800	$41,700	$54,000
Chardonnay	**Total**		**$480,600**	**$431,475**	**$58,575**
	White	**Total**	**$480,600**	**$431,475**	**$58,575**

Row subtotals On

Apply settings to

Per row level   Off

Row level
All

Rows

Show subtotal   On

Subtotal label
Total

Position
Top

Figure 3-7. *Moving the subtotals to the top of each group*

The data is now much clearer to analyze and perhaps we should leave it there. We now have a matrix that shows only subtotals for the SALESPERSON column and doesn't show the column grand totals. We have also removed the stepped layout and moved the row totals and subtotals to the top of each group.

However, perhaps you would like to display the subtotals with a specific format. To do this, on the **Row subtotals** card and the **Rows** subcard, you must ensure that the **Per row level** dropdown is set to "All." Only then will you be able to use the **Values** subcard that presents you with formatting options. To format the row grand total, use the **Row grand total** card; see Figure 3-8.

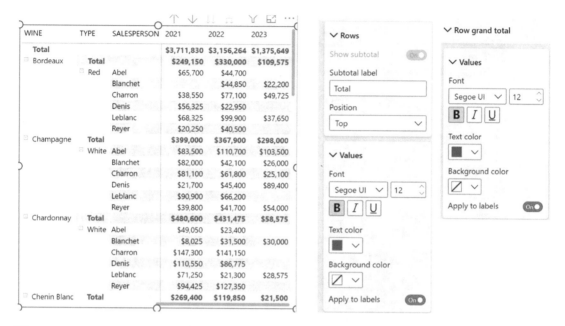

Figure 3-8. *Formatting subtotals*

All these options are also applicable to column-level subtotals and grand totals. Consider the Matrices in Figure 3-9. Here we have put the WINE COUNTRY and SUPPLIER fields in the Rows bucket and the YEAR and QTR fields in the Columns bucket of the matrix. Note that to expand down on the column labels, you must select "Columns" from the **Drill on** dropdown in the visual header. The matrix looks cluttered with subtotals. We only want to show the grand totals for the years. You can see that turning off the subtotals for the QTR column is done in the same way as for row subtotals.

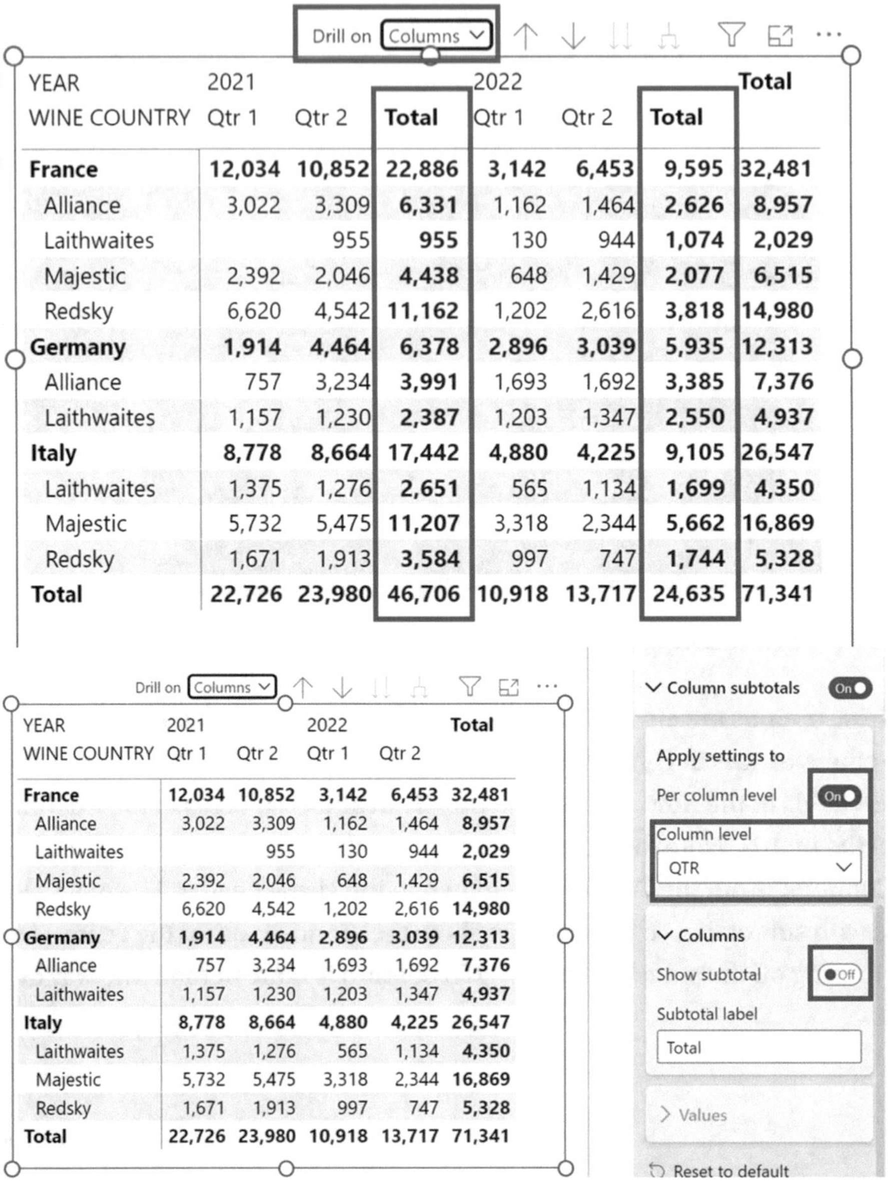

Drill on [Columns ∨] ↑ ↓ ↓↓ ⊥ ▽ ⟐ ⋯

YEAR	2021			2022			Total
WINE COUNTRY	Qtr 1	Qtr 2	Total	Qtr 1	Qtr 2	Total	
France	**12,034**	**10,852**	**22,886**	**3,142**	**6,453**	**9,595**	**32,481**
Alliance	3,022	3,309	**6,331**	1,162	1,464	**2,626**	8,957
Laithwaites		955	**955**	130	944	**1,074**	2,029
Majestic	2,392	2,046	**4,438**	648	1,429	**2,077**	6,515
Redsky	6,620	4,542	**11,162**	1,202	2,616	**3,818**	14,980
Germany	**1,914**	**4,464**	**6,378**	**2,896**	**3,039**	**5,935**	**12,313**
Alliance	757	3,234	**3,991**	1,693	1,692	**3,385**	7,376
Laithwaites	1,157	1,230	**2,387**	1,203	1,347	**2,550**	4,937
Italy	**8,778**	**8,664**	**17,442**	**4,880**	**4,225**	**9,105**	**26,547**
Laithwaites	1,375	1,276	**2,651**	565	1,134	**1,699**	4,350
Majestic	5,732	5,475	**11,207**	3,318	2,344	**5,662**	16,869
Redsky	1,671	1,913	**3,584**	997	747	**1,744**	5,328
Total	**22,726**	**23,980**	**46,706**	**10,918**	**13,717**	**24,635**	**71,341**

Drill on [Columns ∨] ↑ ↓ ↓↓ ⊥ ▽ ⟐ ⋯

YEAR	2021		2022		Total
WINE COUNTRY	Qtr 1	Qtr 2	Qtr 1	Qtr 2	
France	**12,034**	**10,852**	**3,142**	**6,453**	**32,481**
Alliance	3,022	3,309	1,162	1,464	8,957
Laithwaites		955	130	944	2,029
Majestic	2,392	2,046	648	1,429	6,515
Redsky	6,620	4,542	1,202	2,616	14,980
Germany	**1,914**	**4,464**	**2,896**	**3,039**	**12,313**
Alliance	757	3,234	1,693	1,692	7,376
Laithwaites	1,157	1,230	1,203	1,347	4,937
Italy	**8,778**	**8,664**	**4,880**	**4,225**	**26,547**
Laithwaites	1,375	1,276	565	1,134	4,350
Majestic	5,732	5,475	3,318	2,344	16,869
Redsky	1,671	1,913	997	747	5,328
Total	**22,726**	**23,980**	**10,918**	**13,717**	**71,341**

∨ Column subtotals   On ●

Apply settings to

Per column level   On ●

Column level

[QTR   ∨]

∨ Columns

Show subtotal   ● Off

Subtotal label

[Total]

> Values

↺ Reset to default

Figure 3-9. *Turning off subtotals for columns in the matrix*

It does seem that controlling which subtotals and grand totals show in matrices is a little fiddly to say the least, but we can summarize how it's done by saying that to control subtotals you turn on the **Per row (or column) level** option on the **Apply settings to** card and then select the applicable row or column. Controlling the format of grand totals is done on the **Column grand totals** or **Row grand totals** cards.

Q21. Can I Show Different Labels for Different Subtotals?

The problem with the default subtotal labels is that they often don't describe of what they are a subtotal. Take, for example, Figure 3-10 where at the bottom of the visual we have two subtotals and one grand total. It can be confusing as to which subtotal we are looking at.

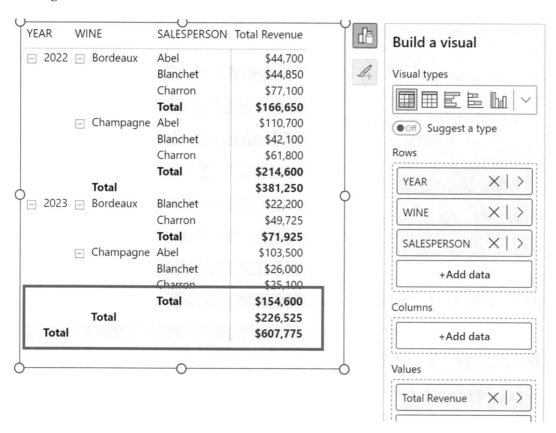

Figure 3-10. *Default subtotal labels can be confusing*

However, if we use the **Row** subtotals card, we can turn on the **Per row level** option, and in the **Row level** dropdown, we can select the row level field for which you would like a different subtotal title, for example, the YEAR field. Then in the **Subtotal label** box, you can type the label you prefer; see Figure 3-11.

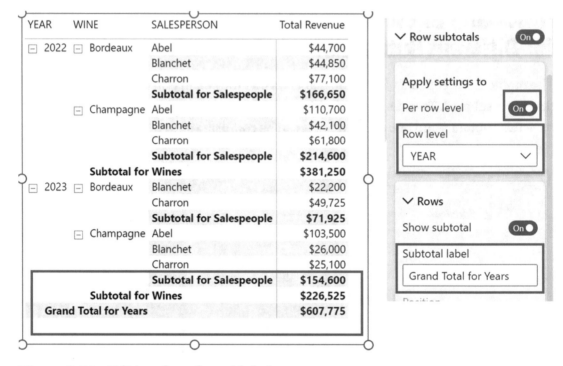

Figure 3-11. *Editing the subtotal labels*

In Q23 we explore changing subtotal calculations whereby each level of subtotal can have a different calculation. You can understand that it would be most important in this scenario to be able to label such subtotals correctly.

Q22. How Do I Sort by Multiple Columns in a Matrix?

To sort a column in a matrix, you can click on the column headers. The problem is that by doing so, you lose any previous sort order. Consider the two matrices in Figure 3-12. The default sort order is alphanumeric on both the WINE and SALESPERSON columns. However, if we sort by the Total Revenue column, we lose this sort order.

WINE	Total Revenue		WINE	Total Revenue ▼
⊟ **Bordeaux**	**$688,725**		⊟ **Champagne**	**$2,285,400**
Abel	$110,400		Blanchet	$432,500
Blanchet	$67,050		Abel	$432,000
Charron	$165,375		Denis	$369,600
Denis	$79,275		Charron	$361,100
Leblanc	$205,875		Reyer	$353,300
Reyer	$60,750		Leblanc	$336,900
⊟ **Champagne**	**$2,285,400**		⊟ **Chardonnay**	**$1,780,275**
Abel	$432,000		Denis	$440,850
Blanchet	$432,500		Reyer	$414,375
Charron	$361,100		Charron	$339,150
Denis	$369,600		Leblanc	$261,975
Leblanc	$336,900		Abel	$201,750
Reyer	$353,300		Blanchet	$122,175
⊟ **Chardonnay**	**$1,780,275**		⊟ **Shiraz**	**$1,522,326**
Abel	$201,750		Denis	$499,980
Total	**$13,698,595**		**Total**	**$13,698,595**

Figure 3-12. Sorting by "Total Revenue" loses the default sort order

Our objective is to retain the sort order of the WINE names and within each wine, to sort by Total Revenue descending as shown in Figure 3-13.

WINE	Total Revenue ▼
⊟ **Bordeaux**	**$688,725**
Leblanc	$205,875
Charron	$165,375
Abel	$110,400
Denis	$79,275
Blanchet	$67,050
Reyer	$60,750
⊟ **Champagne**	**$2,285,400**
Blanchet	$432,500
Abel	$432,000
Denis	$369,600
Charron	$361,100
Reyer	$353,300
Leblanc	$336,900
⊟ **Chardonnay**	**$1,780,275**
Denis	$440,850
Reyer	$414,375

Figure 3-13. *The matrix sorted by WINE and then by Total Revenue*

What we need to do here is to start with a *table* visual. Put the fields into the table visual that are to be put into the matrix. Now click the WINE column header to sort by WINE, and then hold down the SHIFT key and click the Total Revenue column. This will sort the columns correctly; see Figure 3-14.

WINE ▲	SALESPERSON	Total Revenue ▼
Bordeaux	Leblanc	$205,875
Bordeaux	Charron	$165,375
Bordeaux	Abel	$110,400
Bordeaux	Denis	$79,275
Bordeaux	Blanchet	$67,050
Bordeaux	Reyer	$60,750
Champagne	Blanchet	$432,500
Champagne	Abel	$432,000
Champagne	Denis	$369,600
Champagne	Charron	$361,100
Champagne	Reyer	$353,300
Champagne	Leblanc	$336,900
Chardonnay	Denis	$440,850
Chardonnay	Reyer	$414,375
Total		**$13,698,595**

Figure 3-14. *The table sorted by WINE and then by Total Revenue*

Now convert the table to a matrix, moving the fields into the correct buckets as shown in Figure 3-15.

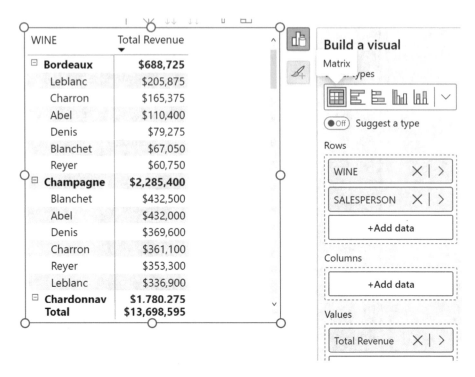

Figure 3-15. *The matrix sorted correctly*

The key here is that you must start with a completely *new* table visual. If the matrix is not sorted correctly, you can't convert the matrix back to a table as the original sort order on the matrix is retained.

Q23. How Can I Control Calculations in the Subtotals?

To answer this question, let's look at the percentages in the "% of Parent" measure in the matrix depicted in Figure 3-16.

YEAR	WINE	Total Revenue	% of Parent	WINE
⊟ 2021	Bordeaux	$249,150	22.07%	■ Bordeaux
	Champagne	$399,000	35.35%	■ Champagne
	Chardonnay	$480,600	42.58%	■ Chardonnay
	Total	**$1,128,750**	**49.99%**	☐ Chenin Blanc
⊟ 2022	Bordeaux	$330,000	29.22%	YEAR
	Champagne	$367,900	32.58%	☐ 2023
	Chardonnay	$431,475	38.20%	■ 2022
	Total	**$1,129,375**	**50.01%**	■ 2021
Total		**$2,258,125**	**100.00%**	☐ 2020

Figure 3-16. *The matrix showing the "% of Parent" calculation*

We understand these percentages because if the same data were in an Excel PivotTable, we would get comparable numbers by showing the values as "**% of Parent Row Total**," as shown in Figure 3-17.

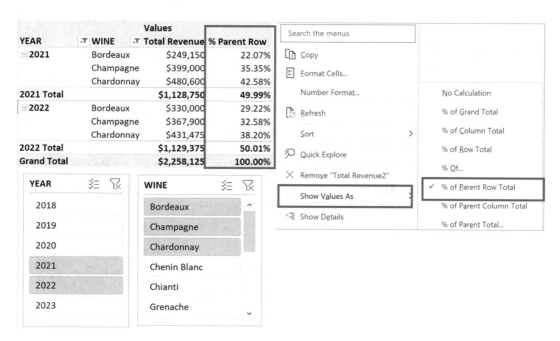

Figure 3-17. *In Excel, you can show values as "% of Parent Row Total"*

You will see that in the matrix visual and in the PivotTable, the wines and years have been filtered by slicers and show sales for three different wines in the years 2021 and 2022. The "% of Parent" column in the matrix and in the PivotTable shows two different percentages.

For each wine, it shows the percentage that wine's sales are of the total for that year for all the selected wines (i.e., in 2022, "Bordeaux" sales are 20.07% of the combined total revenue for "Bordeaux," "Champagne," and "Chardonnay").

On the subtotals for each year however, it shows what the percentage of that year's sales are of the total for both the selected years (i.e., 2021 sales are 49.99% of the total revenue for both years).

As you can see, these percentages are straightforward to see in Excel, but unfortunately, in the matrix visual, there is no option to "show values as percent of parent row total." If you right-click a value in the Values bucket, you will see the options shown in Figure 3-18 and we can, for example, show "Percent of grand total," but this is not the calculation we want.

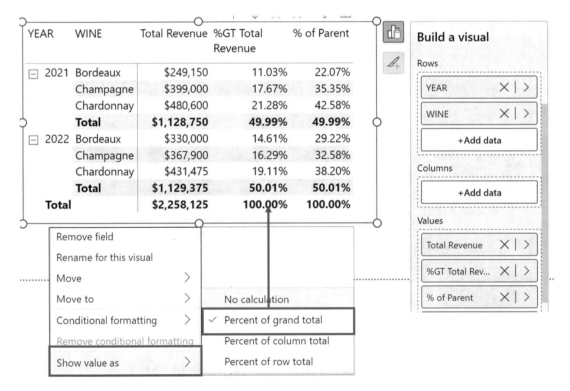

Figure 3-18. *You can show values as "Percent of grand total"*

Therefore, the "% of Parent" calculation you can see in Figures 3-16 and 3-18 was generated using a DAX measure, the details of which we can now explore.

The starting point is to calculate the two different denominators that are required to divide into the Total Revenue. For each wine's sales, it's the value in the subtotal (i.e., for 2021 this will be **$1,128,750** and for 2022 it will be **$1,129,375**), and for the subtotal for each year, it's the grand total of **$2,258,125**.

To calculate the two different denominators, we've used these two DAX measures:

```
All Wines =
CALCULATE ( [Total Revenue], ALLSELECTED ( Products[WINE] ) )

All Years =
CALCULATE ( [Total Revenue], ALLSELECTED ( DateTable[Year] ) )
```

And then used them to divide into the "Total Revenue" measure:-

```
Sales Divided by ALL Wine =
Divide ( [Total Revenue],[All Wines] )

Sales Divided by ALL Years =
Divide ( [Total Revenue],[All Years] )
```

However, you can see the problem with both these measures. In Figure 3-19 we've highlighted where these calculations return an incorrect result. "Sales Divided by ALL Wine" gives us the correct calculation for each wine but not for the subtotal for the year. "Sales Divided by ALL Year" gives us the correct calculation for the yearly subtotals but not for each wine.

YEAR	WINE	Total Revenue	All Wines	All Years	Sales Divided by ALL Wine	Sales Divided by ALL Years		
2021	Bordeaux	$249,150	£1,128,750	£579,150	22.07%	43.02%	**WINE**	
	Champagne	$399,000	£1,128,750	£766,900	35.35%	52.03%	■ Bordeaux	
	Chardonnay	$480,600	£1,128,750	£912,075	42.58%	52.69%	■ Champagne	
	Total	**$1,128,750**	**£1,128,750**	**£2,258,125**	**100.00%**	**49.99%**	■ Chardonnay	
2022	Bordeaux	$330,000	£1,129,375	£579,150	29.22%	56.98%	☐ Chenin Blanc	
	Champagne	$367,900	£1,129,375	£766,900	32.58%	47.97%	**YEAR**	
	Chardonnay	$431,475	£1,129,375	£912,075	38.20%	47.31%	☐ 2023	
	Total	**$1,129,375**	**£1,129,375**	**£2,258,125**	**100.00%**	**50.01%**	■ 2022	
Total		**$2,258,125**	**£2,258,125**	**£2,258,125**	**100.00%**	**100.00%**	■ 2021	
							☐ 2020	

Figure 3-19. *The measures return incorrect results at different levels of the matrix*

To remedy this problem, we need a method to distinguish the different levels in the subtotals of the matrix so we can perform the correct division at the right level, either the YEAR level to divide by "All Years" or the WINE level to divide by "All Wines." To do this, we can use the DAX function HASONEVALUE which returns TRUE when the column you specify has just one value in the current filter. If we create these two measures:

```
Has 1 Value Wine =
HASONEVALUE ( Products[WINE] )
```

```
Has 1 Value Year =
HASONEVALUE( DateTable[Year] )
```

We can see the return values when placed in our matrix; see Figure 3-20.

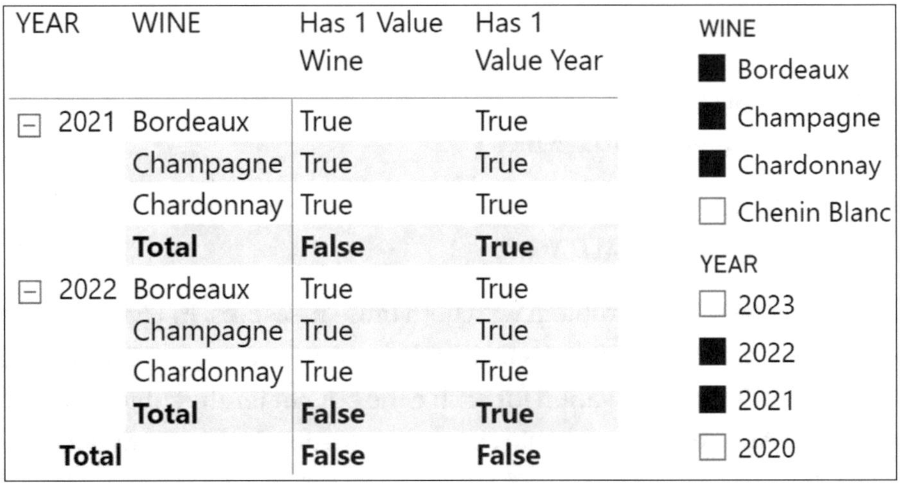

YEAR	WINE	Has 1 Value Wine	Has 1 Value Year
⊟ 2021	Bordeaux	True	True
	Champagne	True	True
	Chardonnay	True	True
	Total	**False**	**True**
⊟ 2022	Bordeaux	True	True
	Champagne	True	True
	Chardonnay	True	True
	Total	**False**	**True**
Total		**False**	**False**

WINE
- ■ Bordeaux
- ■ Champagne
- ■ Chardonnay
- ☐ Chenin Blanc

YEAR
- ☐ 2023
- ■ 2022
- ■ 2021
- ☐ 2020

Figure 3-20. *Using the HASONEVALUE function to test for different subtotals*

Both return True when not being calculated at the subtotal level for the column they reference, and they return False at the subtotal level for the column referenced.

Note   You can't use the ISFILTERED function here because the expression "*=ISFILTERED (Products[WINE])*" will always return TRUE because WINE is being filtered by the slicer. You could, however, use the ISINSCOPE function that would return the same results.

So now we can use a simple IF expression to ensure that we divide by the "All Wines" measure when the evaluation is for each WINE and divide by the "All Years" measure if the evaluation is for the YEAR subtotal or grand total.

```
% of Parent =
IF (
HASONEVALUE ( Products[WINE] ),
DIVIDE ( [Total Revenue], [All Wines] ),
DIVIDE ( [Total Revenue], [All Years] )
)
```

Therefore, using HASONEVALUE in this way allows you to generate different calculations at different levels of subtotals in a matrix.

CHAPTER 4

Date Calculations

There are not many Power BI reports that don't require analysis of data over diverse time frames. By leveraging a date dimension and harnessing the time intelligence functions offered by DAX, calculations involving dates are relatively easy to generate. You don't need to be a DAX guru, for instance, to find the previous year's or month's totals. In this chapter, we've addressed the most frequently asked questions regarding calculations over time that have arisen from working with our clients or from the Power BI community. The information contained in this chapter is certainly something that I'm continually referring to and using in many different situations and scenarios.

Q24. Why Do I Need a Date Table?

If you're a newcomer to Power BI, you may not understand that if you want to analyze yearly, quarterly, or monthly data, or indeed, any other date granularity, you must first generate a date dimension. Once you've done so, you can take advantage of the time intelligence functions within DAX to examine your data from various perspectives, such as years, quarters, months, and days. This unlocks a multitude of insightful opportunities that would otherwise not be possible.

Note For a comprehensive list of DAX time intelligence functions, refer to the DAX function library, `https://learn.microsoft.com/en-us/dax/time-intelligence-functions-dax`.

The problems that arise from the lack of a date table in your data model may not be immediately apparent due to Power BI's automatic generation of an in-memory date hierarchy for each date column in your fact table. This pre-built date hierarchy is depicted in Figure 4-1.

A. Box, *A Power BI Compendium*, https://doi.org/10.1007/978-1-4842-9765-0_4

Figure 4-1. *A date type column with a date hierarchy generated by Power BI*

Note   You can disable the automatic date table generation by going to **Options and settings** on the **File** tab and then **Options**. Under the **Data Load** category for either the Global or Current File, you can find the Time Intelligence "**Auto date/time**" setting.

These date hierarchies enable you to drill down into year, quarter, month, and day. For instance, in Figure 4-2, by using the "**Expand all down one level in the hierarchy**" tool in the line chart visual header, the "SALE DATE" date hierarchy has been used to look into monthly data. This data has been filtered for 2022 by using the Year member of the hierarchy in a slicer.

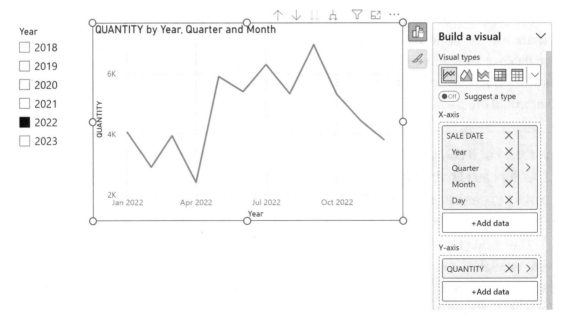

Figure 4-2. *Using the built-in date hierarchy to visualize date data*

You may be wondering, therefore, what are the problems associated with in-built date hierarchies. Firstly, they assume your fiscal year starts in January, which may not align with your actual fiscal year which may begin in April or October. Secondly, visualizing sales comparisons across different years, or other date categories, in a clustered column chart or line chart becomes impossible. Additionally, analyzing data across weeks also presents challenges without a date table. These shortcomings highlight the significance of generating a date dimension and provide just a few compelling reasons for doing so.

Perhaps the next step then is to explore how to generate a date dimension. There are two ways forward with this. You can either

1. Click the **New table** button on the **Modeling** tab, and use the CALENDAR function. This function creates a date table with a specified date range.

    Or

2. Click the **New table** button on the **Modeling** tab, and use the CALENDARAUTO function, which automatically generates a date table based on the earliest and latest dates that can be found in the data model. These dates should be held in a column in the fact table, but specifically it simply looks for a non-calculated column of dates. You can optionally specify your fiscal year month end (e.g., "3" if your fiscal year ends in March). The default is 12.

In Figure 4-3, we have shown how either of these functions can be used when generating the new table.

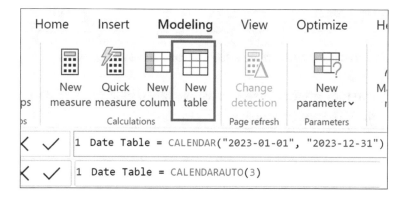

Figure 4-3. *Using the CALENDAR and CALENDARAUTO functions*

Both of these functions will generate a one-column table containing a list of dates. This column is called "Date" by default but you can rename it if required. For instance, in our sample data, this column is called "DATEKEY." You can then create your own calculated columns in this table depending on your requirements. For example, the following DAX expressions would generate Year, Month No., and Month and Qtr columns, as shown in Figure 4-4.

```
YEAR = YEAR('Date Table'[DATEKEY])
QTR = "Qtr "&QUARTER('Date Table'[DATEKEY])
MONTH NO. = MONTH('Date Table'[DATEKEY])
MONTH = FORMAT('Date Table'[DATEKEY],"mmm")
```

DATEKEY	YEAR	QTR	MONTH NO.	MONTH
01/01/2018 00:00:00	2018	Qtr 1	1	Jan
02/01/2018 00:00:00	2018	Qtr 1	1	Jan
03/01/2018 00:00:00	2018	Qtr 1	1	Jan
04/01/2018 00:00:00	2018	Qtr 1	1	Jan
05/01/2018 00:00:00	2018	Qtr 1	1	Jan
06/01/2018 00:00:00	2018	Qtr 1	1	Jan
07/01/2018 00:00:00	2018	Qtr 1	1	Jan
08/01/2018 00:00:00	2018	Qtr 1	1	Jan
09/01/2018 00:00:00	2018	Qtr 1	1	Jan
10/01/2018 00:00:00	2018	Qtr 1	1	Jan
11/01/2018 00:00:00	2018	Qtr 1	1	Jan
12/01/2018 00:00:00	2018	Qtr 1	1	Jan

Figure 4-4. *Generating columns in your custom date dimension*

Sometimes, you may prefer to use an existing date table from an external source, such as Excel or an SQL Server database.

The column containing consecutive dates, which covers the time span of data in your fact table, is the only mandatory column within a date dimension. The remaining columns serve to group and categorize these dates, and their selection can be arbitrary. Typically, you would establish columns for fiscal years, quarters, months (including month names and numbers), and week numbers. To ensure accurate sorting of month names, it's crucial to include both the month name and number.

To guarantee accurate date-based calculations, you must designate your date dimension as a date table. You can achieve this by clicking the **Mark as a date table** button found in the **Table tools** tab. Then, from the dropdown menu, select the column that contains the list of sequential dates; see Figure 4-5.

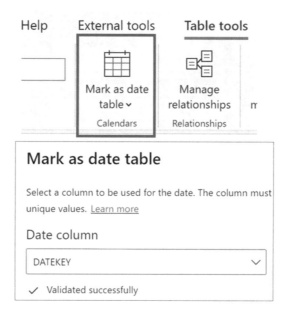

Figure 4-5. *Mark your date dimension as a date table*

To establish a relationship between your date dimension and your fact table, you can use the column containing the list of sequential dates as the linking field. This will generate a many-to-one relationship, as illustrated in Figure 4-6.

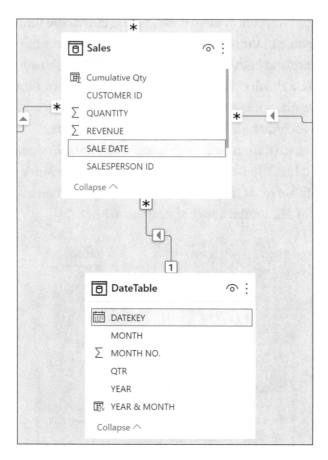

Figure 4-6. *The date dimension is related to the fact table in a many-to-one relationship*

The month names in your date table will be sorted alphanumerically by default when put into a visual. To sort these names correctly, you must use the **Sort by column** button on the **column tools** tab, and, in our example, we would sort MONTH by the MONTH NO.; see Figure 4-7.

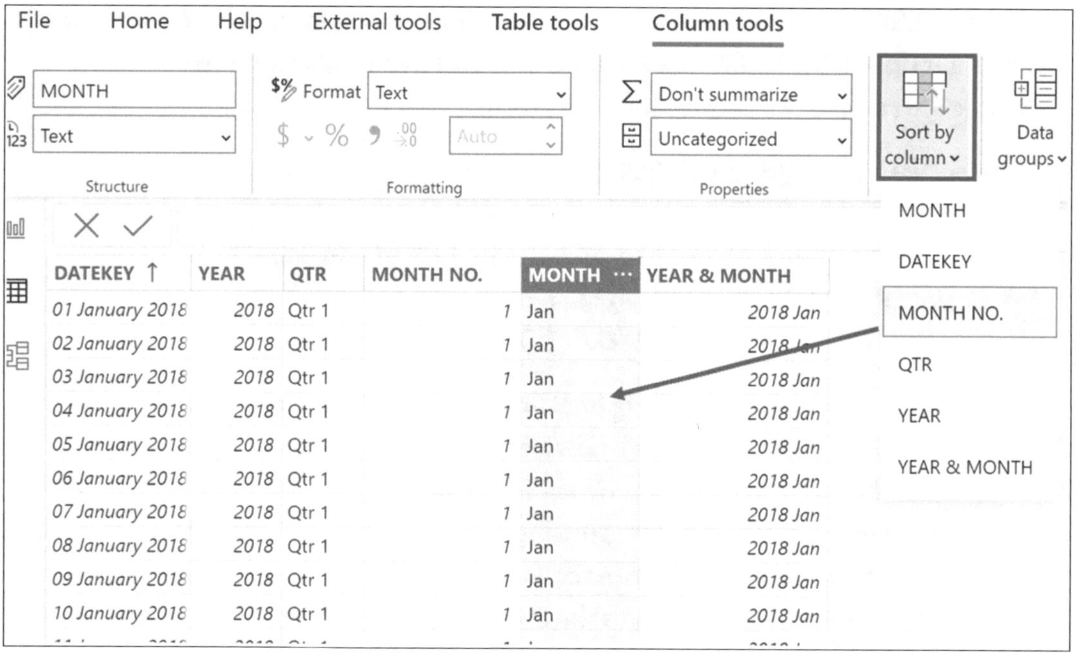

Figure 4-7. *Use the "Sort by column" option to sort the month names*

Once you have incorporated a date table into your data model, you can leverage the capabilities of the time intelligence functions. These functions offer valuable insights into your data, and the answers to the following questions show only some of the possibilities they can unlock.

Q25. How Can I Calculate Rolling Annual Totals and Averages?

This question is a popular one because rolling totals or averages can be more insightful than simple yearly totals or averages. For example, if you're analyzing this year's June sales, you might want to find sales since June last year, not since January this year. To answer this question, we must utilize two functions: DATESINPERIOD and LASTDATE. First, let's compute the rolling annual total revenue:

```
Rolling Annual Total Revenue =
CALCULATE ( [Total Revenue],
    DATESINPERIOD ( DateTable[DATEKEY],
        LASTDATE ( DateTable[DATEKEY] ) , -1 , YEAR ) )
```

The DATESINPERIOD function filters dates commencing from the last date in the current filter (found by the LASTDATE function) and tracing back one year.

This is the code for the rolling annual average:

```
Rolling Annual Average Total Revenue =
CALCULATE (
    AVERAGEX( VALUES ( DateTable[MONTH] ) ,[Total Revenue]),
    DATESINPERIOD ( DateTable[DATEKEY],
        LASTDATE ( DateTable[DATEKEY] ) , -1 , YEAR ) )
```

The difference between the rolling total and the rolling average is that we must determine the average monthly total by dividing by the correct number of rolling months, rather than a simple division by 12. To address this, we can use the VALUES function to generate a virtual table comprising the month names in the MONTH column of the date dimension that are visible in the filter generated by the DATESINPERIOD expression. AVERAGEX then calculates the average total revenue for these months.

To remove evaluations for future dates, use the Filters pane to filter out Total Revenue values that are not blank. The results for both rolling totals can be seen in Figure 4-8.

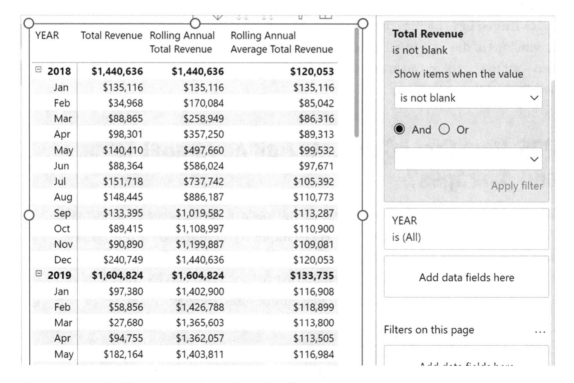

YEAR	Total Revenue	Rolling Annual Total Revenue	Rolling Annual Average Total Revenue
⊟ 2018	$1,440,636	$1,440,636	$120,053
Jan	$135,116	$135,116	$135,116
Feb	$34,968	$170,084	$85,042
Mar	$88,865	$258,949	$86,316
Apr	$98,301	$357,250	$89,313
May	$140,410	$497,660	$99,532
Jun	$88,364	$586,024	$97,671
Jul	$151,718	$737,742	$105,392
Aug	$148,445	$886,187	$110,773
Sep	$133,395	$1,019,582	$113,287
Oct	$89,415	$1,108,997	$110,900
Nov	$90,890	$1,199,887	$109,081
Dec	$240,749	$1,440,636	$120,053
⊟ 2019	$1,604,824	$1,604,824	$133,735
Jan	$97,380	$1,402,900	$116,908
Feb	$58,856	$1,426,788	$118,899
Mar	$27,680	$1,365,603	$113,800
Apr	$94,755	$1,362,057	$113,505
May	$182,164	$1,403,811	$116,984

Total Revenue
is not blank

Show items when the value

is not blank ⌄

● And ○ Or

⌄

Apply filter

YEAR
is (All)

Add data fields here

Filters on this page   ...

Figure 4-8. *Rolling annual totals and rolling averages*

In most time intelligence functions such as DATESINPERIOD and DATESBETWEEN, the "dates" argument is always supplied by the column of consecutive dates in your date table; in our examples, this is the DATEKEY column. This is why the existence of a date dimension is so integral to these date calculations.

Q26. How Do I Show Previous N Months Values from a Selected Month?

This, again, is a question that crops up regularly within the Power BI community. How do I select a month in a slicer and then visualize the next five months' sales from that date? Or how do I show the previous seven months' sales? You can see the solution to this question in Figure 4-9. Here, the data has been visualized in a column chart, and, by using selections in slicers, we are showing the previous four months' sales revenue from January 2022.

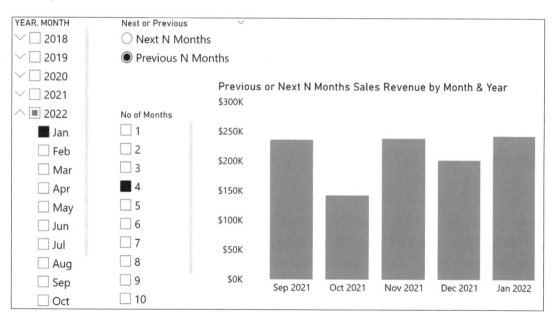

Figure 4-9. *A column chart showing the previous four months' sales from Jan 2022*

Let's now resolve how the column chart in Figure 4-9 was generated. First, you must create the month and year labels to display on the X-axis of the column chart. To do this, create a calculated column in your date table using this DAX expression:

```
Month & Year = STARTOFMONTH ( DateTable[DATEKEY] )
```

This will group your dates by the first of each month, and you can then use the **Format** box on the **Column tools** tab to format the date as "mmm yyyy" as shown in Figure 4-10.

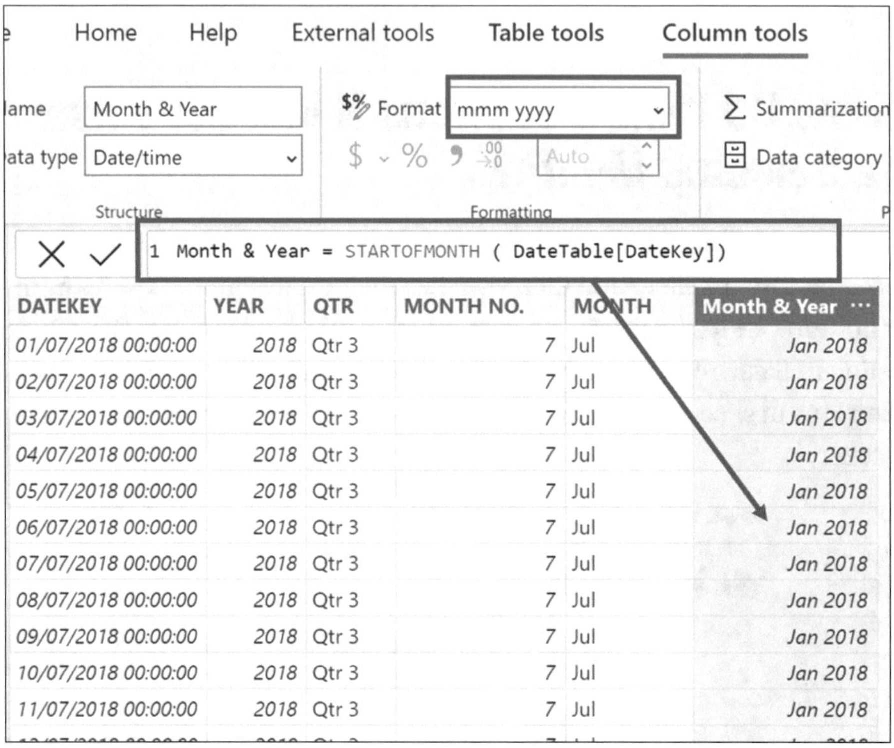

Figure 4-10. *Create a calculated column for "Month & Year"*

So that you can select the base month from which to calculate the previous or next months, you will need a slicer, populated with year and month names. However, we *can't* use the YEAR column and the MONTH column from the DateTable dimension in the slicer because it would always filter the column chart by the selected month. Because of this, we must create a duplicate DateTable from which to select these columns. To make a copy of your date dimension, use the **New table** button on the **Modeling** tab, and into the formula bar enter the DAX expression as shown in Figure 4-11.

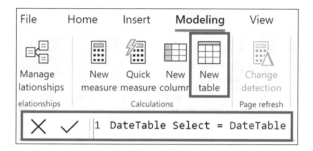

Figure 4-11. *Use the New table button to duplicate the date dimension*

In our example, the duplicate table is called "DateTable Select." This table must not be related to any other tables in the Data Model. Also, you must sort the MONTH column by the MONTH NO. column (see Q24 on sorting month name columns in a date dimension). Now we can put a slicer on the canvas that uses the YEAR and MONTH columns from DateTable Select so we can select the base year and month. This slicer is named YEAR, MONTH in Figure 4-9.

We will need a method to harvest the user choices of either "next" or "previous" and how many months are required. You can do this using two more slicers. To generate the data for these slicers, use the **Enter data** option on the **Home** tab. These are the two tables used in our example:

1. The "**Select Next or Previous**" table comprises two columns "Next or Previous" and "Value."

2. The "**Select N Months**" table contains a single column, "No of Months."

These tables are shown in Figure 4-12.

"Select Next or Previous" table   "Select N Months" table

Next or Previous	▼	Value	▼
Next N Months		1	
Previous N Months		2	

No of Months	▼
1	
2	
3	
4	
5	
6	
7	
8	
9	
10	
11	
12	

Figure 4-12. *Create these two tables using the Enter data button*

We can now put the corresponding two slicers on the canvas populated with the following columns:

1. "Next or Previous" column from the **Select Next or Previous** table. Because the calculation of the DAX measure depends on a selection in this slicer, it makes sense to set it to single select using the **Options** on the **Slicer settings** card in the **Format** pane; see Figure 4-13.

2. "No of Months" column from the **Select N Months** table.

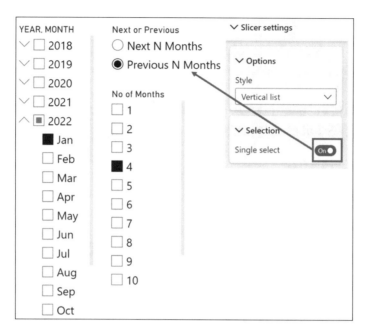

Figure 4-13. *Adding the slicers to the canvas and using Single select*

Now for the challenging part! We must create this DAX measure that will calculate the Total Revenue for the selected months:

```
Previous or Next N months =
VAR MyPreviousDate =
    FIRSTDATE (
        DATEADD (
            'DateTable Select'[DateKey],
            SELECTEDVALUE ( 'Select N Months'[No of Months] )
                * -1,  MONTH )
    )
VAR MyNextDate =
    LASTDATE (
        DATEADD (
            'DateTable Select'[DateKey],
            SELECTEDVALUE ( 'Select N Months'[No of Months] ),
            MONTH
        )
    )
```

```
VAR NextMths =
    SUMX (
        FILTER (
        DateTable,
        DateTable[DateKey] >= MIN ( 'DateTable Select'[DateKey] )
                && DateTable[DateKey] <= MyNextDate
        ),
        [Total Revenue]
    )
VAR PrevMths =
    SUMX (
        FILTER (
            DateTable,
            DateTable[DateKey] >= MyPreviousDate
            && DateTable[DateKey] <=
                MAX ( 'DateTable Select'[DateKey] )
        ),
        [Total Revenue]
    )
RETURN
    SWITCH (
        SELECTEDVALUE ( 'Select Next or Previous'[Value] ),
        1, NextMths,
        2, PrevMths
    )
```

In the evaluation of this measure, it's important to understand that the month filtered on the DateTable will be generated by the month selected in the YEAR, MONTH slicer, and it's the use of DAX variables that drive this calculation, as follows:

- **MyPreviousDate** – harvests the number of months returned by the "Select N Months" slicer and multiplies the value by -1 to return a negative number. This value is used by the DATEADD function to move the dates in the DateKey column of "DateTable Select" backward by that number of months. The FIRSTDATE function then finds the first of these dates. For example, if "**Jan 2022**" is selected in the YEAR, MONTH slicer, "**Previous N Months**" is selected in

the "Next or Previous" slicer, and "**4**" is the number of months, this
will move the dates back to September 2021 and FIRSTDATE (and,
therefore, this variable) will return 1 September 2021.

- **MyNextDate** – harvests the number of months returned by the
  "Select N Months" slicer that is used by the DATEADD function to
  move dates in the DateKey column of "DateTable Select" forward by
  that number of months. The LASTDATE function then finds the last
  date returned by DATEADD.

- **NextMths** – filters the DateTable, starting with the earliest date
  returned by "DateTable Select" (i.e., the month selected in the
  YEAR, MONTH slicer) and ending with the date returned by the
  "MyNextDate" variable. The SUMX function then calculates the Total
  Revenue for these dates.

- **PrevMths** – filters the DateTable, starting with the date returned by
  the "MyPreviousDate" variable, for example, 1 September 2021, and
  ending with the last date returned by "DateTable Select" (i.e., the
  month selected in the YEAR, MONTH slicer, e.g., 31 January 2022).
  The SUMX function then calculates the Total Revenue for these dates.

The measure then returns "NextMths" or "PrevMths" values depending on the
selection in the "Select Next or Previous" slicer.

Now we can construct the column chart. Use the "Year & Month" column from the
DateTable on the X-axis. If the "Year & Month" column presents itself as a date hierarchy
(which it probably will), remove this hierarchy by right-clicking the column name in
the X-axis bucket to select the column and not the hierarchy. The "Previous or Next N
Months" measure goes into the Y-axis bucket; see Figure 4-14.

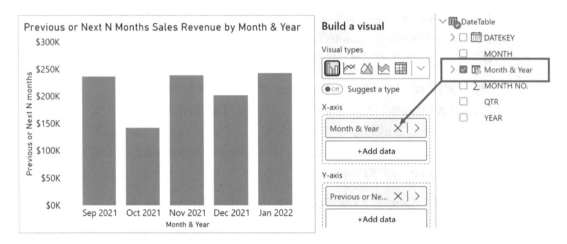

Figure 4-14. *Put the "Month & Year" column from the DateTable into the X-axis and the "Previous or Next N Months" measure into the Y-axis*

The takeaway from the answer to this question is that slicers that use columns from dimensions are always going to filter values in the fact table and therefore place a filter on visuals. If you want to harvest values that can be used as filters on the fact table without filtering the visual, you must create slicers that use columns from unrelated tables.

Q27. How Do I Calculate Cumulative Totals?

It depends where you want to calculate the cumulative totals, whether it's in a calculated column or in a measure. Let's start by showing cumulative values in a calculated column in the Sales table. This is the DAX that will do this job:

```
Cumulative Qty =
VAR MyDate = Sales[SALE DATE]
RETURN
    CALCULATE (
        SUM ( Sales[QUANTITY] ),
        FILTER ( Sales, Sales[SALE DATE] <= MyDate )
    )
```

You can see the results of this expression in Figure 4-15.

```
1  Cumulative Qty =
2  VAR mydate = Sales[SALE DATE]
3  RETURN
4      CALCULATE (
5          SUM ( Sales[QUANTITY] ),
6          FILTER ( Sales, Sales[SALE DATE] <= mydate )
7      )
8
```

SALE DATE	WINESALES NO ↑	SALESPERSON ID	CUSTOMER ID	WINE ID	QUANTITY	REVENUE	Cumulative Qty ⋯
02 January 2018	1	4	18	2	108	$10,800	240
02 January 2018	2	1	72	12	132	$6,600	240
03 January 2018	3	5	78	2	90	$9,000	330
06 January 2018	4	5	40	3	216	$16,200	837
06 January 2018	5	6	3	13	291	$22,698	837
09 January 2018	6	1	12	8	159	$4,770	1188
09 January 2018	7	4	54	7	192	$7,680	1188
10 January 2018	8	3	18	2	106	$10,600	1294
14 January 2018	9	1	35	7	226	$9,040	1520
19 January 2018	10	4	20	9	227	$8,853	1747
22 January 2018	11	3	31	11	227	$10,215	1974
26 January 2018	12	3	11	8	138	$4,140	2335
26 January 2018	13	4	71	7	223	$8,920	2335
27 January 2018	14	1	82	10	140	$5,600	2475
01 February 2018	15	3	26	8	116	$3,480	2591

Figure 4-15. *Calculating cumulative values in a calculated column*

The SALE DATE column does not have to be sorted ascending for the calculation to work. If you want cumulative values for individual categories, for example, for each customer, this would be the expression you would need:

```
Cumulative Qty =
VAR MyDate = Sales[SALE DATE]
VAR MyCustomer = Sales[CUSTOMER ID]
RETURN
    CALCULATE (
        SUM ( Sales[QUANTITY] ),
        FILTER ( Sales, Sales[SALE DATE] <= MyDate &&
            Sales[CUSTOMER ID] = MyCustomer ))
```

Again, you can see this DAX expression in a calculated column of the Sales table in Figure 4-16. Just as this expression does not depend on any sort order in the Sales table, the CUSTOMER ID column does not have to be filtered.

```
   X  ✓      1  Cumulative Qty =
              2  VAR mydate = Sales[SALE DATE]
              3  VAR mycustomer = Sales[CUSTOMER ID]
              4  RETURN
              5      CALCULATE (
              6          SUM ( Sales[QUANTITY] ),
              7          FILTER ( Sales, Sales[SALE DATE] <= mydate && Sales[CUSTOMER ID] = mycustomer )
              8      )
              9
             10
```

SALE DATE ↑	WINESALES NO	SALESPERSON ID	CUSTOMER ID ▽	WINE ID	QUANTITY	REVENUE	Cumulative Qty ···
10 September 2019	216	6	1	11	259	$11,65	259
30 October 2019	243	3	1	3	184	$13,80	443
25 November 2019	251	5	1	2	220	$22,00	663
19 December 2019	257	5	1	8	112	$3,36	775
31 December 2019	260	1	1	10	348	$13,92	1123
23 July 2020	369	3	1	13	248	$19,34	1371
01 October 2020	403	2	1	12	380	$19,00	1751
07 December 2020	439	3	1	11	311	$13,99	2062
10 December 2020	443	2	1	13	449	$35,02	2511
18 May 2021	589	3	1	4	334	$28,39	2845

Figure 4-16. *Calculating cumulative values for customers in a calculated column*

However, most cumulative totals are calculated in the context of a DAX measure so that they can be used in a visual. For example, in Figure 4-17 we have calculated year-to-date values for the QUANTITY column in a table visual.

YEAR	Total Qty	Year To Date Qty
⊟ **2021**	**72,666**	**72,666**
Jan	10,574	10,574
Feb	4,754	15,328
Mar	7,398	22,726
Apr	9,026	31,752
May	8,743	40,495
Jun	6,211	46,706
Jul	4,578	51,284
Aug	5,949	57,233
Sep	4,400	61,633
Oct	2,728	64,361
Nov	4,437	68,798
Dec	3,868	72,666

Figure 4-17. *Year-to-date values in a table visual*

To calculate cumulative year-to-date values in a measure, you can use the DATESYTD function as follows:

```
Year To Date Quantity =
CALCULATE ( [Total Qty] ,
    DATESYTD ( DateTable[DATEKEY] ) )
```

You have the flexibility to set your own year-end date in the second argument of this function. To avoid any complications related to the date locale, you can provide your year-end date in the format "yyyy-mm-dd" as shown here:

```
Year To Date Qty =
CALCULATE ( [Total Qty] ,
    DATESYTD ( DateTable[DATEKEY], "2021-03-31")
)
```

You can use any year in the date format because the year part is ignored in the evaluation of the expression.

However, perhaps you must calculate cumulative total-to-date values. There are two ways that this can be done: using the DATESBETWEEN function or using a direct filter on the DateTable. This is the measure for the former option:

```
Cumulative Total =
CALCULATE ( [Total Qty] ,
    DATESBETWEEN ( DateTable[DATEKEY], 0 ,
        LASTDATE ( DateTable[DATEKEY] )
    )
)
```

Figure 4-18 shows the results of using this expression. This function generates a table containing dates that lie within the specified range, determined by the second and third arguments. The DATESBETWEEN function's start date is set to zero, ensuring it corresponds to the earliest value in the DATEKEY column. The end date is determined using the LASTDATE function, which corresponds to the last date of the month found in any row of the table or matrix visual, or the final date of a month filtered in a slicer or the Filters pane.

YEAR	Total Qty	Cumulative Total
⊟ **2021**	**72,666**	**170,679**
Jan	10,574	108,587
Feb	4,754	113,341
Mar	7,398	120,739
Apr	9,026	129,765
May	8,743	138,508
Jun	6,211	144,719
Jul	4,578	149,297
Aug	5,949	155,246
Sep	4,400	159,646
Oct	2,728	162,374
Nov	4,437	166,811
Dec	3,868	170,679
⊟ **2022**	**56,788**	**227,467**
Jan	4,065	174,744
Feb	2,909	177,653
Mar	3,944	181,597
Apr	2,409	184,006
May	5,903	189,909
Jun	5.405	195.314

Figure 4-18. *The cumulative total quantities*

The alternative method of creating a cumulative total is to place a direct filter on the DATEKEY column of the DateTable as follows:

```
Cumulative Total =
CALCULATE (
    [Total Qty],
        DateTable[DATEKEY] <= MAX ( DateTable[DATEKEY] ) )
```

Here, each date in the DATEKEY column is compared to the date obtained from the expression *"MAX (DateTable[DATEKEY])."* This expression is responsible for identifying the last date within the present filter. For instance, if we evaluate "May 2021," it will return **31 May 2021**, filtering out all dates before or on this particular date.

The second approach to cumulative totals will return exactly the same values as you can see in Figure 4-18, and in this respect, it's a personal choice as to which you want to use. However, you can appreciate that by using the DATESBETWEEN function you can analyze values between any two dates. See also QU64 "How Can I Calculate Cumulative Values Using Power Query?"

Q28. How Do I Find the Difference Between Two Dates?

You can simply subtract dates in one column from the dates in another. These can be columns in Table view using a calculated column or two measures that sit in columns of a table or matrix visual. However, we must nest the dates inside the INT function to obtain a value in days, rather than a date. For example, we can calculate the first and last transaction dates for our customers with the following two measures:

```
Date of First Transaction = MIN(Sales[SALE DATE])
```

```
Date of Last Transaction = MAX(Sales[SALE DATE])
```

Now we can find the difference in days between these dates:

```
Days Difference =
INT([Date of Last Transaction]) -
          INT([Date of First Transaction])
```

Just like Excel, DAX also provides the DATEDIFF function that allows you to calculate the difference between weeks, months, years, etc. For instance, to calculate the difference in months in a measure, you would use this code:

```
Months Difference =
DATEDIFF ( [Date of First Transaction],
             [Date of Last Transaction], MONTH )
```

In Figure 4-19, you can see these measures evaluated for each customer.

CUSTOMER NAME	Date of First Transaction ▲	Date of Last Transaction	Days Difference	Months Difference
Beaverton & Co	02/01/2018	05/02/2023	1,860	61
Melbourne Ltd	02/01/2018	26/09/2022	1,728	56
Miyagi & Co	03/01/2018	30/04/2023	1,943	63
Cape Canaveral Ltd	06/01/2018	11/02/2023	1,862	61
Fremont & Sons	06/01/2018	11/04/2023	1,921	63
Brooklyn & Co	09/01/2018	27/03/2023	1,903	62
El Cajon & Sons	09/01/2018	04/04/2023	1,911	63
Eilenburg Ltd	14/01/2018	30/03/2023	1,901	62
Clifton Ltd	19/01/2018	29/04/2023	1,926	63
Issaquah & Co	22/01/2018	28/02/2023	1,863	61
Canoga Park Ltd	26/01/2018	04/01/2023	1,804	60
Hawthorne Bros	26/01/2018	24/12/2022	1,793	59
East Orange & Co	27/01/2018	03/10/2022	1,710	57
Old Savbrook Ltd	01/02/2018	06/01/2023	1.800	59

Figure 4-19. *Calculating days and months between two dates*

However, you may want to find the days difference between dates that sit in different rows of the fact table using a calculated column. Consider Figure 4-20 where we have calculated the number of days between sales transactions, as shown in the "PREV DATE DAYS DIFF" calculated column.

SALE DATE	WINESALES NO	PREVIOUS DATE	PREV DATE DAYS DIFF
02 January 2018	1		
02 January 2018	2	02 January 2018	0
03 January 2018	3	02 January 2018	1
06 January 2018	4	03 January 2018	3
06 January 2018	5	06 January 2018	0
09 January 2018	6	06 January 2018	3
09 January 2018	7	09 January 2018	0
10 January 2018	8	09 January 2018	1
14 January 2018	9	10 January 2018	4
19 January 2018	10	14 January 2018	5
22 January 2018	11	19 January 2018	3
26 January 2018	12	22 January 2018	4
26 January 2018	13	26 January 2018	0
27 January 2018	14	26 January 2018	1
01 February 2018	15	27 January 2018	5

Figure 4-20. *The "PREV DATE DAYS DIFF" expression in a calculated column*

To perform this calculation, we need to locate the SALE DATE from the previous row. The challenge when working with calculated columns, where the row context confines expressions to use values from the current row, is how to calculate values from a different row. This is done by providing an index that will number each transaction sequentially (we will use the WINESALES NO column but you can generate index values using Power Query). If we then filter the sales table to match rows where the index is less than the index of the current row, we can find the latest date in these filtered rows, which will be the date previous to the date of the current row. This is the expression in the calculated column that will do this job:

```
PREVIOUS DATE =
VAR MyIndex = Sales[WINESALES NO]
VAR PreviousDate =
    CALCULATE (
        MAX ( Sales[SALE DATE] ),
        FILTER ( Sales, Sales[WINESALES NO] < MyIndex ) )
```

```
RETURN
PreviousDate
```

Now we can calculate the days difference:

```
PREV DATE DAYS DIFF =
VAR MyIndex = Sales[WINESALES NO]
VAR PreviousDate =
    CALCULATE (
        MAX ( Sales[SALE DATE] ),
        FILTER ( Sales, Sales[WINESALES NO] < MyIndex ) )
RETURN
 IF( PreviousDate, INT (Sales[SALE DATE]) - INT ( PreviousDate) )
```

The "PREV DATE DAYS DIFF" column subtracts the current row's SALE DATE from the date generated by the "PreviousDate" variable and returns a date.

If you want to calculate previous transaction dates in the context of a DAX measure, you can do this using the same logic that was used in the calculated column. That is simply finding the maximum date that is less than the date in the current filter:

```
Previous Sales Date =
IF (
    [Total Revenue],
    CALCULATE (
        MAX ( Sales[SALE DATE] ),
        Sales[SALE DATE] < SELECTEDVALUE ( Sales[SALE DATE] ) ) )
```

Note that by using the IF function, this code is evaluated only for dates that have a Total Revenue value.

We can now find the difference in days between the previous and current transaction:

```
Days Difference Measure 2 =
DATEDIFF([Previous Sales Date],SELECTEDVALUE(Sales[SALE DATE]),DAY)
```

If you need to retrieve the value associated with the previous transaction date, you can use the LASTNONBLANKVALUE function as follows:

```
Previous Sales Value =
IF (
    [Total Revenue],
CALCULATE (
    LASTNONBLANKVALUE ( Sales[SALE DATE], [Total Revenue] ),
    Sales[SALE DATE] < SELECTEDVALUE ( Sales[SALE DATE] )))
```

These measures are illustrated in Figure 4-21.

SALE DATE	Total Revenue	Previous Sales Date	Days Difference Measure 2	Previous Sales Value
02 January 2018	$17,400			
03 January 2018	$9,000	02 January 2018	1	$17,400
06 January 2018	$38,898	03 January 2018	3	$9,000
09 January 2018	$12,450	06 January 2018	3	$38,898
10 January 2018	$10,600	09 January 2018	1	$12,450
14 January 2018	$9,040	10 January 2018	4	$10,600
19 January 2018	$8,853	14 January 2018	5	$9,040
22 January 2018	$10,215	19 January 2018	3	$8,853
26 January 2018	$13,060	22 January 2018	4	$10,215
27 January 2018	$5,600	26 January 2018	1	$13,060
01 February 2018	$3,480	27 January 2018	5	$5,600
04 February 2018	$4,590	01 February 2018	3	$3,480
08 February 2018	$3,360	04 February 2018	4	$4,590
10 February 2018	$3,360	08 February 2018	2	$3,360
16 February 2018	$6,840	10 February 2018	6	$3,360
27 February 2018	$13,338	16 February 2018	11	$6,840

Figure 4-21. *Calculating the difference in values between consecutive transactions using a DAX measure*

Therefore, the answer to the question "How do I find the difference between two dates" is that it depends! It depends on whether you want to find the difference between two dates that sit in columns where you can simply subtract the values or use the DATEDIFF function. However, if you want to find the difference between previous dates that sit in rows of the same column, you must find the maximum date that is less than the date in question.

Q29. How Can I Toggle Between Different Date Metrics?

So you want to create dynamic date-based metrics that automatically adjust depending on the user's selection in a slicer? Typically, these might be toggling between year-to-date, quarter-to-date, and month-to-date values. We can do this easily in Power BI using time intelligence functions in conjunction with a parameter table that drives values for the slicer.

We must first generate the parameter table that will determine the user's selection. Use the **Enter data** button on the **Home** tab, and type labels that correspond to the choice of metrics required. An example of such labels is shown in Figure 4-22.

	Slicer Selection	+
1	YTD	
2	QTD	
3	MTD	
+		

Figure 4-22. *An example of labels to be used in the choice of date metrics*

You can call this table, for example, "Date Metrics" and save and load it. You can now create a measure called "Dynamic Sales" using the following DAX formula:

```
Dynamic Sales =
SWITCH (
SELECTEDVALUE ( 'Date Metrics'[Slicer Selection] ),
"YTD",
CALCULATE ([Total Revenue],DATESYTD( DateTable[DATEKEY] )),
"QTD",
CALCULATE ([Total Revenue],DATESQTD( DateTable[DATEKEY] )),
"MTD",
CALCULATE ([Total Revenue],DATESMTD( DateTable[DATEKEY] )),
[Total Revenue]
)
```

By using a slicer with options like "YTD," "QTD," and "MTD," your report users can seamlessly switch between year-to-date, quarter-to-date, and month-to-date calculations; see Figure 4-23.

Slicer Selection	YEAR	Total Revenue	Dynamic Sales
■ MTD	2018	$1,440,636	$240,749
☐ QTD	Qtr 1	$258,949	$88,865
☐ YTD	Jan	$135,116	$135,116
	02 January 2018	$17,400	$17,400
	03 January 2018	$9,000	$26,400
	06 January 2018	$38,898	$65,298
	09 January 2018	$12,450	$77,748
	10 January 2018	$10,600	$88,348
	14 January 2018	$9,040	$97,388
	19 January 2018	$8,853	$106,241
	22 January 2018	$10,215	$116,456
	26 January 2018	$13,060	$129,516
	27 January 2018	$5,600	$135,116
	Feb	$34,968	$34,968
	01 February	$3,480	$3,480
	04 February	$4,590	$8,070
	08 February	$3,360	$11,430
	10 February	$3,360	$14,790
	16 February	$6,840	$21,630
	27 February	$13,338	$34,968

Figure 4-23. *Using a slicer to browse by "MTD," "QTD," "YTD"*

Of course, it doesn't have to be total-to-date values used in this way. You could instead use the PREVIOUSMOUTH, PREVIOUSQUARTER, and PREVIOUSYEAR functions to toggle values accordingly, or indeed, many other time intelligence functions could be utilized dynamically in conjunction with parameter tables and slicers.

Q30. How Can I Show Open Items in Any Month?

This is a question that has many scenarios. Often it's projects that, for any given month, you want to know which projects were still current, or it could be helpdesk tickets that you want to know which tickets were ongoing. On the other hand, you may have been asked in what months employees were working for you. In Figure 4-24, you can see how we could visualize these scenarios. In the matrix on the left, the "ID" column could represent projects, helpdesk tickets, or employees. We then have a YEAR column and "Start" and "End" dates. Under each month's column, a black dot shows if the ID is still current in that month. This then corresponds to the summary matrix on the right.

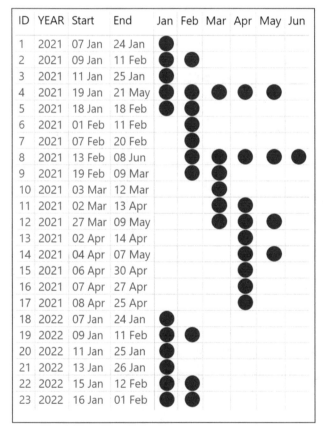

Figure 4-24. *A matrix visual showing open items for any given month with a visual showing the summary*

Let's first look at the source data for these visuals. In Figure 4-25, you can see that the Open Items table holds the Start and End dates for the IDs. The month names used in the column labels of the matrix are taken from the MONTH column in the DateTable. The YEAR row labels are also taken from the YEAR column in the DateTable. Note that the Open Items table is not related to any other tables in the Data Model.

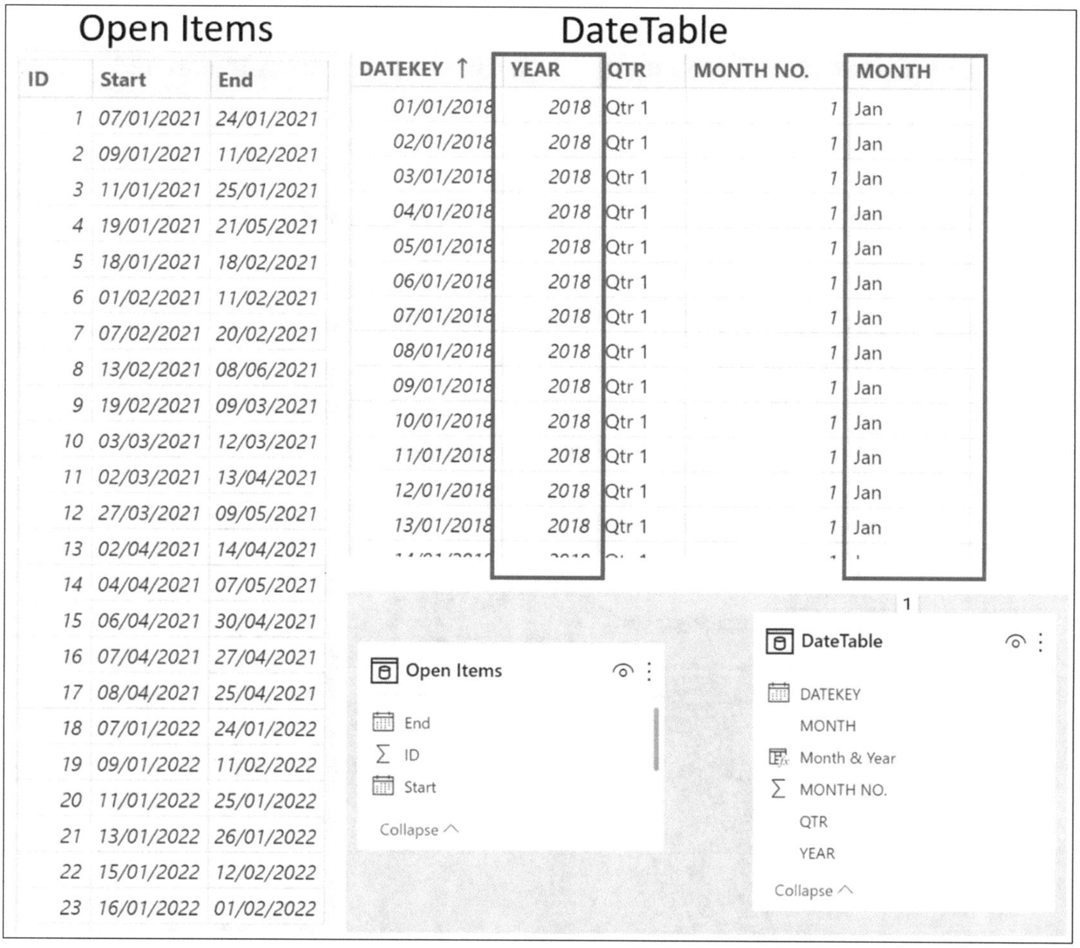

Figure 4-25. *The source data used in Figure 4-24. The Open Items table is not related to any other table*

Now we need two measures, the first for the matrix showing the black circles:

```
No of Open =
CALCULATE (
    COUNTROWS ( 'Open Items' ),
```

```
MONTH ( 'Open Items'[Start] ) <= MAX ( DateTable[Month No.] )
    && MONTH ( 'Open Items'[End] ) >=
                            MAX ( DateTable[Month No.] )
    && YEAR ( 'Open Items'[End] ) = MAX ( DateTable[Year] ) )
```

This measure uses filters on columns in the Open Items table to filter the corresponding dates in the DateTable.

The second measure is for the summary matrix that generates subtotals and totals:

```
No of Open Total =
SUMX (
  SUMMARIZE ( DateTable, DateTable[YEAR], DateTable[MONTH NO.] ),
    [No of Open] )
```

Because we have put the YEAR from the DateTable into the rows of the matrix, we don't need to show the year in the Start and End dates in the Open Items table. Therefore, the next step is to format the Start and End columns as "dd mmm" using the **Format** dropdown on the **column tools** tab; see Figure 4-26.

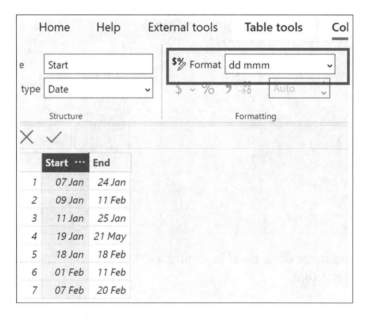

Figure 4-26. *Format the Start and End columns as "dd mmm"*

Now we can build the matrix that shows open items in any month as shown in Figure 4-27. On the **Format** pane, you can change the **Style presets** to "None," and you can also use the **Grid** card to show gridlines.

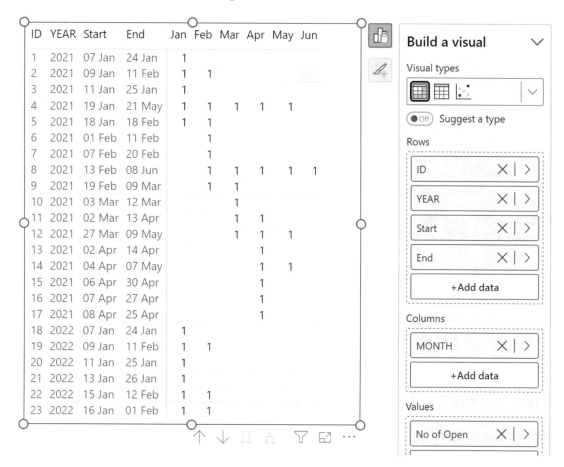

Figure 4-27. *The Matrix showing the open items in any month*

Of course, having black dots rather than showing 1's in the matrix looks a little more engaging! To represent the data in this way, use the conditional formatting option on the **Cell elements** card. Turn on the **Icons**, and in the conditional formatting dialog, ensure that in the **Icon layout** dropdown, you select "Icon only." Then set your rule accordingly to show the back dot icon (or any other icon of your choosing); see Figure 4-28.

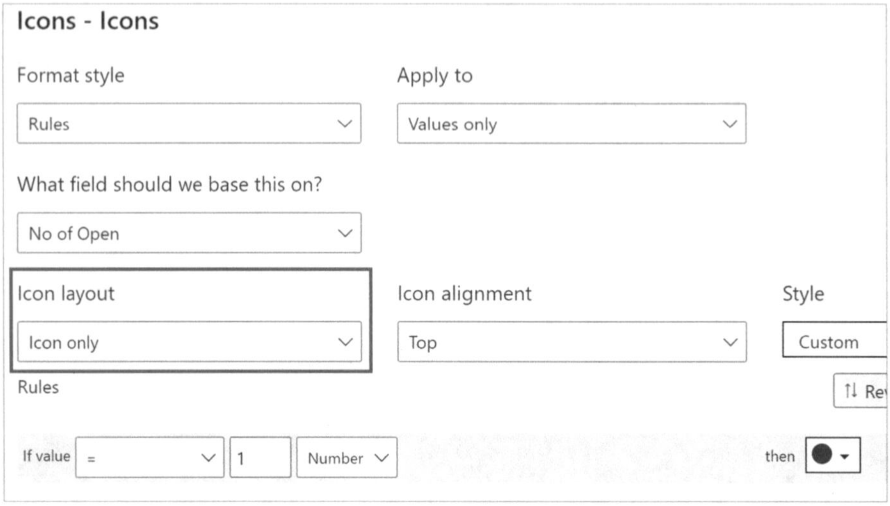

Figure 4-28. *Use the "Icon only" option in the conditional formatting dialog to represent the data as an icon*

The key element of this analysis is using the YEAR and MONTH columns in the DateTable that are not related to the dates in the Open Items table. In this way, these columns are being used as parameters against which to filter the relevant rows in the Open Items table so they can then be counted.

Q31. Is There Weekly "Time Intelligence" in DAX?

Well, no there's not. There are no time intelligence functions for calculations on weekly data in DAX. The problem with weeks is that they don't fall into a date hierarchy. Weeks can span consecutive months and consecutive quarters. In other words, weeks are not subsets of months and quarters. If you want to perform calculations across weeks, you must do this by constructing the necessary DAX yourself.

However, the starting point for weekly calculations is to ensure you have a date dimension in your data model (see Q24, "Why Do I Need a Date Table?"). In your date table, you must generate columns for week and weekday numbers as follows:

```
WEEK = WEEKNUM(DateTable[DATEKEY],2)
```

This will start your week on a Monday. Use "1" in the second argument if your week starts on a Sunday.

```
WEEKDAY = WEEKDAY(DateTable[DATEKEY],2)
```

In the WEEKDAY function, we have used "2" in the second argument to assign "1" to Monday. You can see these calculated columns in Figure 4-29.

DATEKEY	YEAR	QTR	MONTH NO.	MONTH	YEAR & MONTH	WEEK	WEEKDAY
01 January 2018	2018	Qtr 1	1	Jan	Jan 2018	1	1
02 January 2018	2018	Qtr 1	1	Jan	Jan 2018	1	2
03 January 2018	2018	Qtr 1	1	Jan	Jan 2018	1	3
04 January 2018	2018	Qtr 1	1	Jan	Jan 2018	1	4
05 January 2018	2018	Qtr 1	1	Jan	Jan 2018	1	5
06 January 2018	2018	Qtr 1	1	Jan	Jan 2018	1	6
07 January 2018	2018	Qtr 1	1	Jan	Jan 2018	1	7
08 January 2018	2018	Qtr 1	1	Jan	Jan 2018	2	1
09 January 2018	2018	Qtr 1	1	Jan	Jan 2018	2	2
10 January 2018	2018	Qtr 1	1	Jan	Jan 2018	2	3
11 January 2018	2018	Qtr 1	1	Jan	Jan 2018	2	4
12 January 2018	2018	Qtr 1	1	Jan	Jan 2018	2	5
13 January 2018	2018	Qtr 1	1	Jan	Jan 2018	2	6
14 January 2018	2018	Qtr 1	1	Jan	Jan 2018	2	7
15 January 2018	2018	Qtr 1	1	Jan	Jan 2018	3	1
16 January 2018	2018	Qtr 1	1	Jan	Jan 2018	3	2
17 January 2018	2018	Qtr 1	1	Jan	Jan 2018	3	3

Figure 4-29. *The WEEK and WEEKDAY calculated columns in the DateTable*

Now we are ready to author some weekly calculations. Let's start with calculating the previous week's sales revenue:

```
Previous Week =
VAR MaxWeek =
    CALCULATE ( MAX ( DateTable[Week] ), ALL ( DateTable[Week] ) )
RETURN
    IF (
        SELECTEDVALUE ( DateTable[Week] ) = 1,
        CALCULATE (
            [Total Revenue],
            FILTER (
                ALL ( DateTable ),
                DateTable[Week] = MaxWeek
                    && DateTable[Year]
                        = SELECTEDVALUE ( DateTable[Year] ) - 1
        )),
```

```
CALCULATE([Total Revenue],
    ALL(DateTable[Week]),
    DateTable[Week]=SELECTEDVALUE(DateTable[Week])-1
))
```

In the "Previous Week" measure, for week 1 we must calculate total revenue for week 52 or 53 of the previous year. Therefore, we must find the maximum week number, that is, either 52 or 53, and store it in the variable "MaxWeek." If the evaluation of the measure is for week 1, then total revenue will be calculated for "MaxWeek" in the previous year. For any other weeks, the total revenue is calculated for the previous week; see Figure 4-30.

YEAR	WEEK	Total Revenue	Previous Week
⊟ 2021	44		$61,480
	45	$46,380	
	46	$110,208	$46,380
	47	$30,810	$110,208
	48	$51,399	$30,810
	49	$79,266	$51,399
	50	$14,340	$79,266
	51	$54,445	$14,340
	52	$22,950	$54,445
	53	$30,935	$22,950
	Total	**$3,711,830**	
⊟ 2022	1		$30,935
	2	$111,530	
	3	$4,080	$111,530
	4	$48,575	$4,080
	5	$78,690	$48,575

Figure 4-30. *The "Previous Week" measure*

Now let's build the calculation for the same week in the previous year where there are two considerations:

1. There will be values for future weeks which we must eliminate. We can do this by finding the minimum date in the current filter, and then only compute values if this date is less than or equal to the last date in the Sales table.

2. The Subtotal and Total rows must show the total revenue for the previous year or years. This will be a different calculation than that for the same week in the previous year. In QU23, "How Can I Control Calculations in Subtotals?" we learned that we can use the HASONEVALUE function to accomplish this task.

This is the measure we can now author:

```
Same Wk Previous Year =
VAR CurrentDate =
    MIN ( DateTable[DATEKEY] )
VAR MaxDate =
    CALCULATE ( MAX ( Sales[SALE DATE] ), ALL ( Sales ) )
VAR ValuesToShow = CurrentDate <= MaxDate
RETURN
    IF (
        ValuesToShow,
        IF (
            HASONEVALUE ( DateTable[WEEK] ),
            CALCULATE (
                [Total Revenue],
                ALL ( DateTable[WEEK] ),
                DateTable[Week] = SELECTEDVALUE ( DateTable[Week] ),
                DateTable[Year]
                    = SELECTEDVALUE ( DateTable[Year] ) - 1
            ),
            CALCULATE ( [Total Revenue], DATEADD ( DateTable[DATEKEY], -1,
            YEAR ) ) ) )
```

You can see the results of the "Same Wk Previous Yr" measure in Figure 4-31.

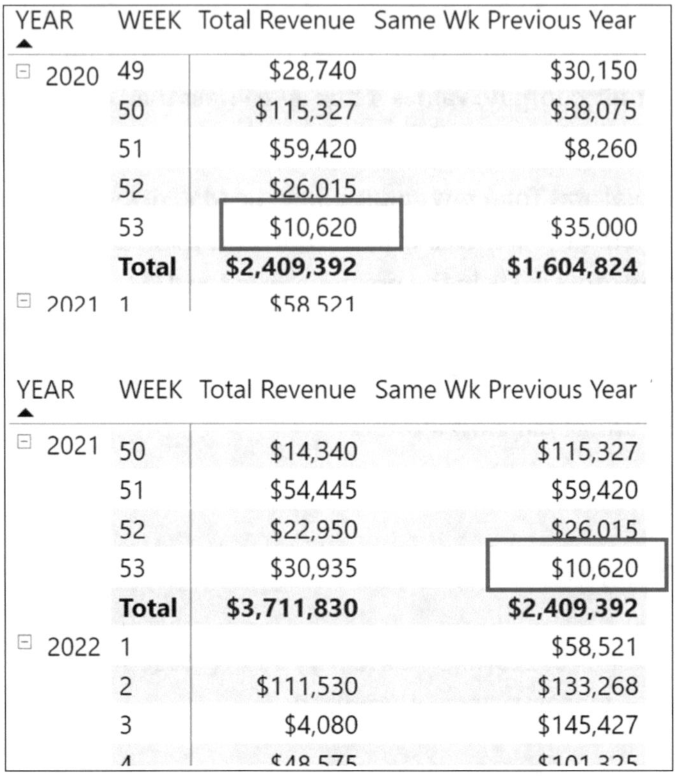

YEAR	WEEK	Total Revenue	Same Wk Previous Year
⊟ 2020	49	$28,740	$30,150
	50	$115,327	$38,075
	51	$59,420	$8,260
	52	$26,015	
	53	$10,620	$35,000
	Total	**$2,409,392**	**$1,604,824**
⊟ 2021	1	$58,521	

YEAR	WEEK	Total Revenue	Same Wk Previous Year
⊟ 2021	50	$14,340	$115,327
	51	$54,445	$59,420
	52	$22,950	$26,015
	53	$30,935	$10,620
	Total	**$3,711,830**	**$2,409,392**
⊟ 2022	1		$58,521
	2	$111,530	$133,268
	3	$4,080	$145,427
	4	$48,575	$101,325

Figure 4-31. *The "Same Wk Previous Year" measure*

Lastly, we will calculate week-to-date total sales revenue where again we must ensure that we don't get evaluations for future weeks and that the Total row returns a value:

```
Week to Date Sales =
VAR CurrentDate =
    MIN ( DateTable[DATEKEY] )
VAR MaxDate =
    CALCULATE ( MAX ( Sales[SALE DATE] ), ALL ( Sales ) )
VAR ValuesToShow = ( CurrentDate <= MaxDate )
RETURN
    IF (
        ValuesToShow,
        IF (
            HASONEVALUE ( 'DateTable'[WEEK] ),
            CALCULATE (
```

```
        [Total Revenue],
        ALL ( DateTable ),
        DateTable[WEEKDAY] <= MAX ( DateTable[WEEKDAY] ),
        DateTable[WEEK] = SELECTEDVALUE ( DateTable[WEEK] ),
        DateTable[YEAR] = SELECTEDVALUE ( 'DateTable'[YEAR] )
    ),
    [Total Revenue]
  )
)
```

In this measure, by removing the filter from the DateTable, we can filter any rows where the weekday number is less than the weekday number in the current filter and the week and year are equal to the week and year in the current filter. The measure is illustrated in Figure 4-32.

YEAR ▼	WEEK	DATEKEY	Total Revenue	Week to Date Sales
⊟ 2023	⊟ 16	15/04/2023	$4,200	$59,325
		16/04/2023		$59,325
		Total	**$59,325**	**$59,325**
	⊟ 17	19/04/2023	$24,780	$24,780
		20/04/2023	$9,984	$34,764
		21/04/2023		$34,764
		22/04/2023		$34,764
		23/04/2023	$11,840	$46,604
		Total	**$46,604**	**$46,604**
	⊟ 18	24/04/2023	$26,000	$26,000
		25/04/2023		$26,000
		26/04/2023	$6,870	$32,870
		27/04/2023	$41,184	$74,054
		28/04/2023		$74,054
		29/04/2023	$25,230	$99,284
		30/04/2023	$26,180	$125,464
		Total	**$125,464**	**$125,464**
	Total		**$1,375,649**	**$1,375,649**
Total			**$1,375,649**	**$1,375,649**

Figure 4-32. *The "Week to Date Sales" measure in a matrix*

Unfortunately, Power BI does not give us any easy ways to calculate weekly data. However, with the help of DAX and a date dimension, you can perform these calculations yourself.

Q32. How Can I Make Comparisons Between Years?

You could, for instance, create two visuals with each visual interacting with a different slicer. You can use the **Edit interactions** button on the **Format tab** to control which visual interacts with each slicer. One visual could show a selected year's data, and the other visual could show a comparison year; see Figure 4-33.

YEAR	WINE	Total Revenue
☐ 2018	Bordeaux	$249,150
☐ 2019	Champagne	$399,000
☐ 2020	Chardonnay	$480,600
	Chenin Blanc	$269,400
■ 2021	Chianti	$207,080
☐ 2022	Grenache	$293,610
☐ 2023	Malbec	$287,130
	Merlot	$211,536
	Piesporter	$200,700
	Pinot Grigio	$176,490
	Rioja	$312,300
Edit	Sauvignon Blanc	$186,240
interactions	Shiraz	$438,594
	Total	**$3,711,830**

YEAR	WINE	Total Revenue
☐ 2018	Bordeaux	$330,000
☐ 2019	Champagne	$367,900
☐ 2020	Chardonnay	$431,475
	Chenin Blanc	$119,850
☐ 2021	Chianti	$191,920
■ 2022	Grenache	$133,620
☐ 2023	Malbec	$378,760
	Merlot	$180,375
	Piesporter	$181,170
	Pinot Grigio	$35,310
	Rioja	$326,430
	Sauvignon Blanc	$132,120
	Shiraz	$347,334
	Total	**$3,156,264**

Figure 4-33. *Using two visuals interacting with two different slicers*

However, we could perform the comparison using just one visual, where it would be easier to compare values as illustrated in Figure 4-34.

YEAR ∨	WINE	Total Revenue	Total Revenue Compare	YEAR COMPARE
☐ 2018	Bordeaux	$249,150	$330,000	☐ 2018
☐ 2019	Champagne	$399,000	$367,900	☐ 2019
☐ 2020	Chardonnay	$480,600	$431,475	☐ 2020
	Chenin Blanc	$269,400	$119,850	
☑ 2021	Chianti	$207,080	$191,920	☐ 2021
☐ 2022	Grenache	$293,610	$133,620	☑ 2022
☐ 2023	Malbec	$287,130	$378,760	☐ 2023
	Merlot	$211,536	$180,375	
	Piesporter	$200,700	$181,170	
	Pinot Grigio	$176,490	$35,310	
	Rioja	$312,300	$326,430	
	Sauvignon Blanc	$186,240	$132,120	
	Shiraz	$438,594	$347,334	
	Total	**$3,711,830**	**$3,156,264**	

Figure 4-34. Using one visual interacting with two different slicers

Should you opt for this approach to compare two years, you must first create a comparison DateTable. This can be accomplished using either DAX or Power Query. Ensure that the comparison DateTable is related to the Sales fact table, in our instance, we used the SALE DATE and DATEKEY columns as the connecting fields. Next, it's crucial to modify this relationship to mark it as inactive. To achieve this, go to the **Edit Relationship** dialog, and check off the option **Make this relationship active** which is shown in Figure 4-35.

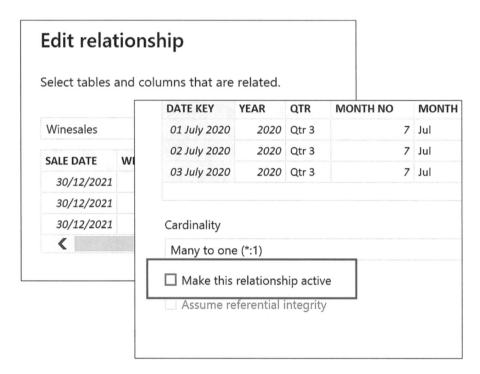

Figure 4-35. *Making a relationship inactive*

In Figure 4-36, you can see how the "DateTable Compare" table is related to the Sales fact table using an inactive relationship.

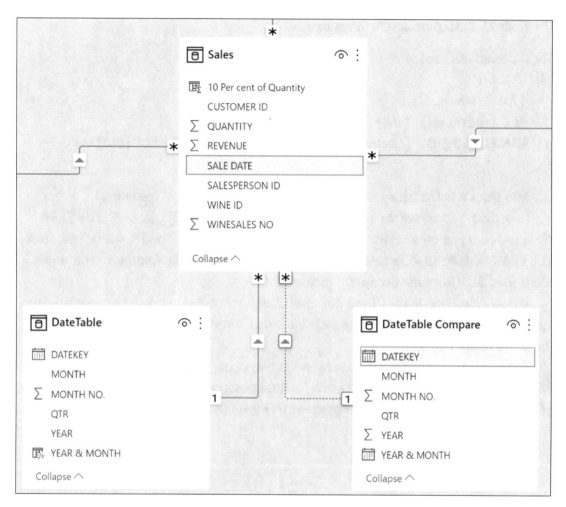

Figure 4-36. *The DateTable Compare table related to the Sales table using an inactive relationship*

To finish the report, create this measure:

```
Total Revenue Compare =
CALCULATE (
    [Total Revenue],
    ALL ( DateTable[YEAR] ),
    USERELATIONSHIP ( Sales[SALE DATE], 'DateTable Compare'[DATEKEY] )
)
```

And place it in the table visual alongside the "Total Revenue" measure.

Now place two slicers on the report canvas, as depicted in Figure 4-36. The YEAR slicer uses the year column from the DateTable and filters the "Total Revenue" measure. The YEAR COMPARE slicer uses the column from the DateTable Compare table and filters the "Total Revenue Compare" measure.

The ALL function in the "Total Revenue Compare" measure is used to clear the filter applied on the YEAR column of the DateTable that comes from the YEAR slicer used by the active relationship.

Therefore, to make comparisons between two years, you can use the USERELATIONSHIP function that activates filters on a comparison date table. You could, of course, also compare months or quarters by introducing slicers, respectively.

CHAPTER 5

Data Modeling

Data modeling may not seem glamorous or exciting, which is why many new Power BI users tend to skip right past it and dive into the more exhilarating task of creating visualizations. But the best visualizations are those that actually deliver something worthwhile, and you'll need a solid data model to get at those powerful insights. However, it still persists that people pay the least attention to designing the data model that will sit at the heart of their report. Often, it will be a case of importing tables, letting Power BI link the tables with default relationships and leaving it at that. Unfortunately, this approach can lead to the creation of problematic many-to-many bi-directional relationships that hinder accurate reporting and calculations. To achieve meaningful results, it's essential to invest time and attention in designing a robust data model that forms the core of your report.

As Power BI trainers, a significant portion of our efforts is dedicated to emphasizing the significance of creating effective data models. Regrettably, this advice goes mostly unheeded, and it's only later that people encounter difficulties in achieving the reports they are eager to generate. When confronted with these challenges, it becomes evident that a re-evaluation of the data model's structure is necessary. In this chapter, we tackle the issues that arise when you finally recognize the imperative need to organize and fine-tune data tables and their relationships in order to achieve your desired reporting goals.

Q33. Why Do I Need a Star Schema?

Before we answer this question, perhaps we should step back and answer the question that naturally precedes this, and that is, "What is a star schema"? Only then can we address the question of why we need one.

A. Box, *A Power BI Compendium*, https://doi.org/10.1007/978-1-4842-9765-0_5

In our sample data, we've imported six tables into our data model as follows:

- **Sales** – records our sales transactions.

- **Products** – records the names and details of the wines we sell.

- **Customers** – who we sell our wines to.

- **SalesPeople** – the people making the sales.

- **Regions** – our customers are grouped into these regions.

- **DateTable** – records every date, starting from the first day of the month when sales start and ending with the last date in the current financial year, categorizing these dates into year, quarter, and month.

Our tables are related in many-to-one relationships which you can see in **Model view** as shown in Figure 5-1.

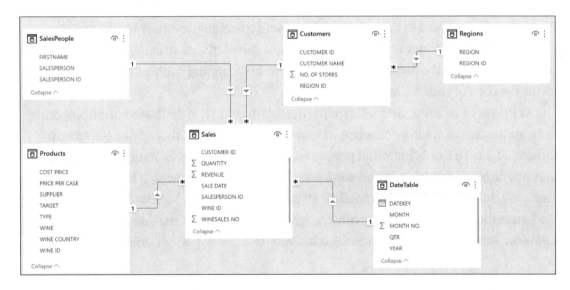

Figure 5-1. *The data model that is used in our examples*

You will observe that the DateTable, SalesPeople, Customers, and Products tables are all related to the Sales table. Notice the "**1**" and "*****" to denote the one side and the many side of the relationship, respectively. The columns used to create the relationships have the same column names in both tables, for example, WINE ID in the Products Table is related to WINE ID in the Sales table. The Regions table is the odd one out in that it's not directly related to the Sales table but *indirectly* related via the Customers table.

> **Note**   If you would like more information on creating relationships between tables in Power BI Desktop, follow this link:
>
> `https://docs.microsoft.com/en-us/power-bi/desktop-create-and-manage-relationships`

In answering the question, what is a star schema, let's start by considering the different types of tables that comprise the model. In a Power BI data model, a table should be either one of two types: either a *fact table* or a *dimension*.

The fact table stores "events." The term "event" is used loosely here to describe activities such as sales, orders, survey results, etc. Fact tables answer the question *what*? That is, what are you analyzing in your report? You can identify the fact table by asking yourself these three questions:

1. Which table holds the data that you want to analyze in your report?

2. If you delete this table, will the remaining tables in the data model still be related between each other?

3. Which table sits on the many side of all the other relationships?

In our data model, the Sales table is our fact table. By definition, fact tables sit on the many side of a many-to-one relationship. Another attribute of the fact table is that the data in it will change frequently and it'll probably have many more rows than a dimension.

Dimension tables, on the other hand, store the descriptions of the entities in your model and sit on the one side of a many-to-one relationship. Dimensions answer the question *how*? That is, how do you want to analyze your data? In our data model, we can analyze our sales by products, salespeople, customers, regions, and by dates using the data in the columns within these tables. The data in dimensions does not necessarily change regularly, and dimensions tend to have fewer rows than fact tables.

There's no table property that you set to configure the table type as a dimension or a fact table. It's determined by which side of the relationship the table sits on. Tables that sit on the "one" side are dimension-type tables while tables that are *only* related on the "many" side are fact tables.

The reason it's so important to distinguish between these two different types of tables is that they support two different types of behavior in the data model, as follows:

- Dimension tables support *grouping* and *filtering*.

- Fact tables support *summarizations*.

This brings us to the concept of a *star schema*. You'll notice in Model view that we've placed the fact table in the middle of the view and arranged our dimensions around the fact table. This arrangement can be described as a star shape, giving the name to the structure. In a perfect star schema, all dimensions are directly related to the fact table; see Figure 5-2.

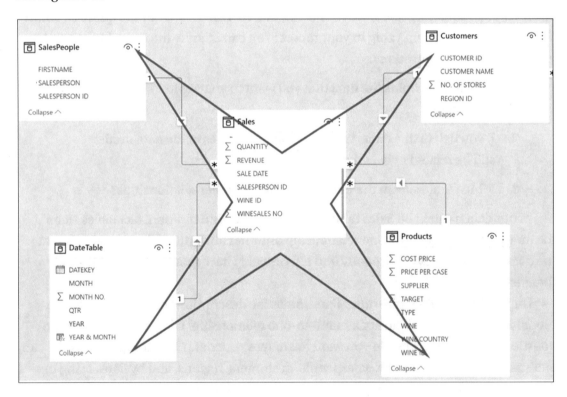

Figure 5-2. *The star schema*

There is an imperfection in our data model because the Regions table is a dimension related to another dimension. Dimensions that are not directly related to a fact table but are indirectly related via dimension tables are described as *snowflake* dimensions. You can imagine that if we had a number of dimensions related to other dimensions in chains outward from the fact table, the schema would more resemble a snowflake.

Now we know what a star schema is, we can now answer the more pressing question: Why do I need one? Many people design their Power BI reports using one big flat table (sometimes referred to as a "BFT"). Others will use a model comprising many tables but no clearly defined fact tables or dimensions, bi-directional filters, and many-to-many relationships. Neither of these is a satisfactory state of affairs. Perhaps we should listen to Marco Russo and Alberto Ferrari who in their blog, "Power BI – Star schema or single table," say:

The reason experienced BI professionals always choose a star schema is because they have failed several times with other models. All those times, the solution to their problems was to build a star schema. You may make one, two, ten errors. After those, you learn to trust star schemas. This is how you become an expert: after tons of errors, you finally learn how to avoid them.[1]

So what is all the fuss about? Where is the value in using the star schema? We're going to focus on just four benefits to be gained by implementing this structure, but there are many more, particularly relating to performance and refresh issues.

Note   For more information about the performance issues between star schemas and flat tables, visit Marco Russo and Alberto Ferrari's blog, "Power BI – Star schema or single table" mentioned in Footnote 1.

The first benefit is that simply finding your way around your data becomes much easier with the star schema. Categorizing your data into dimensions means you know exactly where the data concerning each category is stored. This data can also be easily added to and amended accordingly.

The second benefit is also one of the key aspects of DAX. What newbies to DAX often overlook is that the details of your DAX expressions will be inextricably tied to the structure of the model. The simpler the model, the more straightforward the calculations. For instance, having a date dimension means that you can reap the benefits of the DAX time intelligence functions. Also, DAX measures must often perform calculations on aggregated data (e.g., finding the average of totals), and these calculations are facilitated with the existence of dimensions.

---

[1] Marco Russo and Alberto Ferrari, "Power BI – Star schema or single table," April 2021, www.sqlbi.com/articles/power-bi-star-schema-or-single-table.

There is another property of the star schema that makes DAX expressions easier to author and that is that filters should only flow from dimension tables to fact tables and never in the opposite direction. Using bi-directional filtering is never a good idea as it becomes increasingly difficult to trace filter propagation through the model on the evaluation of DAX measures. We explore this aspect of the data model in more detail in Q38 "Why Doesn't My Measure Work?"

Because data is infinitely variable, the tables in your data model may not be arranged obediently in a perfect star schema. Having multiple fact tables, for instance, isn't necessarily a problem. The thing to bear in mind, however, is that the more your model diverges from a star schema, particularly where you don't have clearly defined fact tables and dimensions and you implement bi-directional filtering, the more you will need complex DAX to manage it.

The third reason to use dimensions and fact tables is that often you must show where there is no data, for instance, showing which customers had no sales last month. This analysis is implicit in the star schema. In Q16, "How Do You Filter Using a DAX Measure?" we explored the option to "show items with no data." This option would not be applicable if the data were sitting in a flat table or did not have the correct relationship to the fact table. But also the fact table will show non-matching values from dimensions related to it. You may have seen the "blank" value showing in visuals. This indicates there are values in the fact table that don't link to any values in dimensions. For example, we may have WINE ID's recorded in the Sales table that don't exist in the Products dimension; see Figure 5-3.

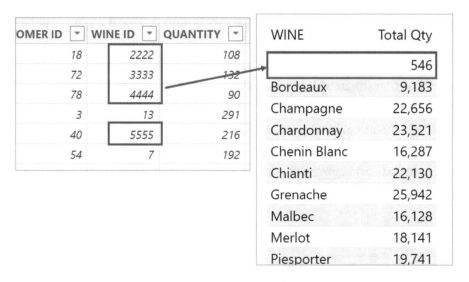

Figure 5-3. *The blank entry shows when values don't match in dimensions*

All three of these reasons to adopt a star schema as your data model are certainly persuasive. However, there is an overriding reason why you should always use dimensions related to fact tables. That reason is that many calculations just don't work when you have ad hoc tables with no clearly defined roles. Let's now explore two examples where this is true.

In our sample data, let's say that our Sales table is a BFT ("big flat table"), and we have no dimensions. In that case, all the customers' data that was stored in the Customers dimension would now be held in the Sales table. For simplicity, we have just moved two fields, the CUSTOMER NAME and NO. OF STORES fields into the Sales table; see Figure 5-4.

DATE	WINESALES NO	SALESPERSON ID	CUSTOMER ID	WINE ID	QUANTITY	REVENUE	CUSTOMER NAME	NO. OF STORES
huary 2018	5	6	3	13	291	$22,698	Cape Canaveral Ltd	17
ruary 2018	20	6	65	13	171	$13,338	Warsaw Ltd	17
March 2018	25	6	10	10	135	$5,400	Lavender Bay Ltd	10
April 2018	33	6	48	7	91	$3,640	Shanghai Ltd	5
April 2018	37	6	59	5	108	$3,240	Branch Ltd	23
April 2018	39	6	70	12	278	$13,900	Runcorn & Co	3
May 2018	44	6	44	13	103	$8,034	Saarbrücken & Co	16
June 2018	58	6	8	3	291	$21,825	Charlottesville & Co	13
July 2018	69	6	79	5	83	$2,490	St. Leonards Ltd	6
July 2018	72	6	19	7	240	$9,600	Townsville Ltd	22
July 2018	74	6	26	9	172	$6,708	Old Saybrook Ltd	12
ugust 2018	80	6	84	12	152	$7,600	Runcorn Ltd	25
ugust 2018	82	6	76	5	217	$6,510	Waterbury Ltd	10
ugust 2018	86	6	27	11	83	$3,735	Plattsburgh Ltd	12
ugust 2018	90	6	84	3	135	$10,125	Runcorn Ltd	25
mber 2018	96	6	25	8	265	$7,950	Chatou & Co	8
mber 2018	102	6	20	11	246	$11,070	Clifton Ltd	15
mber 2018	110	6	5	2	286	$28,600	Snoqualmie & Sons	13
mber 2018	118	6	65	12	282	$14,100	Warsaw Ltd	17
mber 2018	121	6	43	12	262	$13,100	Jacksonville Ltd	1
mber 2018	127	6	71	10	289	$11,560	Canoga Park Ltd	12

Figure 5-4. *The flat Sales table now contains all the Customers' data*

Each customer has stores in which we sell our product, recorded in the NO. OF STORES column. We have been asked to calculate the total number of stores in which we have sold each product. The intuitive calculation would be to simply sum the values in the NO. OF STORES column in the fact table:

```
NO. OF STORES WRONG =
SUM ( Sales[NO. OF STORES] )
```

However, if we take the WINE ID column from the flat Sales table to identify each product, this measure returns incorrect values as shown in Figure 5-5 where we can compare this measure to the correct calculation.

WINE ID	NO. OF STORES CORRECT	NO. OF STORES WRONG
1	244	371
2	777	1,246
3	748	1,318
4	499	771
5	715	1,412
6	566	930
7	590	1,112
8	606	906
9	581	845
10	660	1,098
11	625	1,036
12	561	906
13	527	1,007
Total	**1,112**	**12,958**

Figure 5-5. *Using a flat table returns incorrect results*

Because each customer's data is repeated with each sales transaction, summing the NO. OF STORES values won't work. These values are an attribute of each customer, not of their transactions, and, therefore, this value should be stored in a dimension. In the star schema data model, where we have a Customers dimension, we can hold the NO. OF STORES value in this table; see Figure 5-6.

CUSTOMER ID ↓↑	CUSTOMER NAME ▾	REGION ID ▾	NO. OF STORES ▾
1	Landstuhl Ltd	1800	21
2	Erlangen & Co	800	21
3	Cape Canaveral Ltd	800	17
4	Black Ltd	500	19
5	Snoqualmie & Sons	1200	13
6	Leeds & Co	1000	11
7	Newcastle upon Tyne & Sons	1900	18
8	Charlottesville & Co	300	13
9	Brown & Co	1400	24
10	Lavender Bay Ltd	500	10
11	Hawthorne Bros	1200	17
12	El Cajon & Sons	400	1
13	Sedro Woolley Ltd	1800	23
14	Parkville Ltd	900	11

Figure 5-6. The Customers dimension and the NO. OF STORES column

We can then author this correct calculation:

```
NO. OF STORES CORRECT =
CALCULATE ( SUM( Customers[NO. OF STORES] ), Sales )
```

We examine this measure more closely in Q35, "How Can I Manage Many-to-Many Relationships?" Suffice it to say here, that we are using CALCULATE to programmatically pass the current filter on the Sales fact table onto the Customers dimension.

If this still hasn't persuaded you that star schemas should always be the way forward in Power BI data modeling, let me cite one more example where calculations can go dramatically wrong when passed over flat tables. The problem is that when measures filter multiple columns from the *same* table, the DAX engine, for reasons of efficiency, takes the combination of data in these columns and automatically filters out any combinations where there is no data. This can lead to unexpected results.

To illustrate this, let's take the scenario where we have recorded each wine that our salespeople have sold each year in a flat table (named "FlatTable"). For example, "Leblanc" sold two wines in 2021, "Bordeaux" and "Grenache"; see Figure 5-7.

Year	Wine	Salesperson
2020	Bordeaux	Denis
2021	Merlot	Denis
2021	Bordeaux	Leblanc
2021	Grenache	Leblanc
2021	Merlot	Blanchet
2022	Merlot	Denis
2022	Grenache	Denis
2022	Bordeaux	Leblanc
2022	Grenache	Leblanc

Figure 5-7. *The "FlatTable" table*

We want to count the number of products each salesperson has sold in each year and show this in a matrix. This is the measure that will do this job:

```
No. of Products Sold =
COUNTROWS ( FlatTable )
```

We can see the results of this measure when placed in a matrix; see Figure 5-8.

Salesperson	Year	No. of Products Sold
⊟ Blanchet	2021	1
	Total	**1**
⊟ Denis	2020	1
	2021	1
	2022	2
	Total	**4**
⊟ Leblanc	2021	2
	2022	2
	Total	**4**
Total		**9**

Figure 5-8. *A matrix showing the number of products sold by each salesperson in each year*

We would now like to compare the number of products sold in each year to the total number in all the years for the selected salespeople by using this measure:

```
No. of Products Sold All Years =
CALCULATE ( [No. of Products Sold], ALL ( FlatTable[Year] ) )
```

We can put the "No. of Products Sold" and the "No. of Products Sold All Years" measures into card visuals and now browse the data using two slicers, one for Salesperson and the other for Year. However, problems arise when we start to browse the data by year when considering only salespeople, "Denis" and "Leblanc"; see Figure 5-9. When filtering 2021, the "No. of Products Sold All Years" measure is correct. However, when filtering by 2020, the measure returns an incorrect result of "**4**." It should return "**8**" in both cases as "Denis" and "Leblanc" have sold eight wines between them in all years overall.

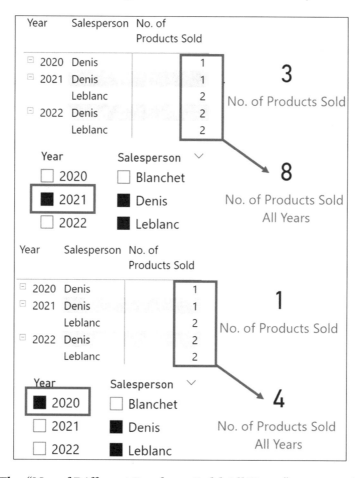

Figure 5-9. *The "No. of Different Products Sold All Years" measure is not correct for 2020*

The predicament lies with the fact that "Leblanc" didn't sell any wine in 2020, and, therefore, the DAX engine removes "Leblanc" from the evaluation because there is no data to show. Therefore, when the ALL function removes the filter from the Year column, "Leblanc" has already been removed, so the measure returns all the rows only for "Denis" who sold four products in all the years.

On the other hand, if we had dimensions for Year, Salesperson, and Wine, filters would be driven by each dimension, and so both "Leblanc" and "Denis" would remain as part of the filter. Figure 5-10 shows what these dimensions and fact table would look like.

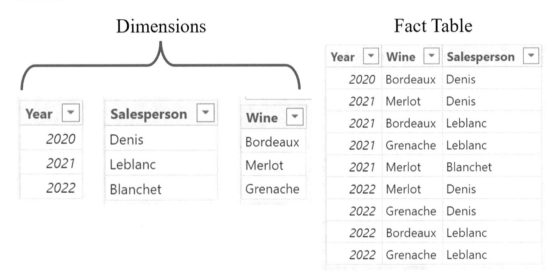

Figure 5-10. *Creating dimensions will pass correct filters to the fact table*

We can now generate a star schema from the dimensions and the fact table; see Figure 5-11.

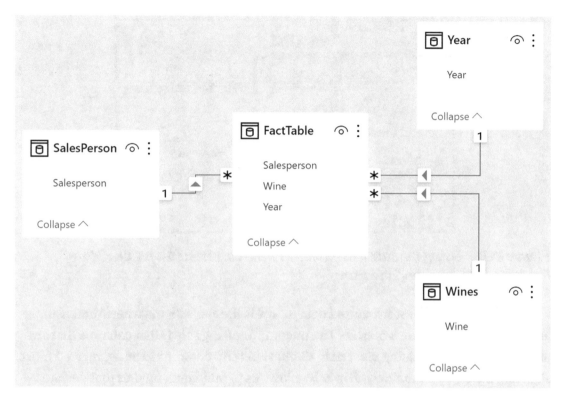

Figure 5-11. *Generating a star schema from the dimensions and the fact table*

We must now make an amendment to the "No. of Products Sold All Years" measure whereby the ALL function is passed over the Year column in the Year dimension:

```
No. of Products Sold All Years =
CALCULATE ( [No. of Products Sold],ALL( Year[Year] ) )
```

When we build the matrix and slicers, the Salesperson and Year fields will come from the dimensions, and the measure now returns the correct result; see Figure 5-12.

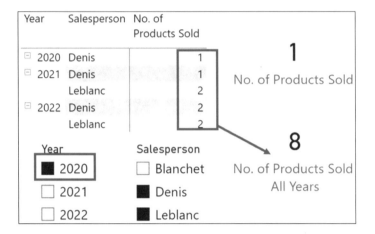

Figure 5-12. *Using the star schema returns the correct result for the "No. of Products Sold All Years" measure*

The simplest and best structured data model is the star schema where dimensions are related directly to the fact table. The uncompromising rule is that columns that are used for filtering and slicing must sit in dimensions and columns to be aggregated must sit in fact tables. Unfortunately, Power BI allows us to build data models that deviate from this rule, but this is never a good idea.

Q34. How Can I Create Dimensions from a Flat Table?

So we have established in Q33, "Why Do I Need a Star Schema?" that we must avoid big flat tables in our data model. The question that must now be asked is, so how do I generate a star schema from a big flat table? How do I generate dimensions when they don't at first exist in the model? In fact, creating dimensions is so simple that there is no excuse for not doing so! However, for this, we must use Power Query which is always the preferred tool for preparing data for a star schema. In Figure 5-13, you can see that we have imported into Power Query a flat Sales table that comprises the WINE column. This column must be extracted into a Products dimension.

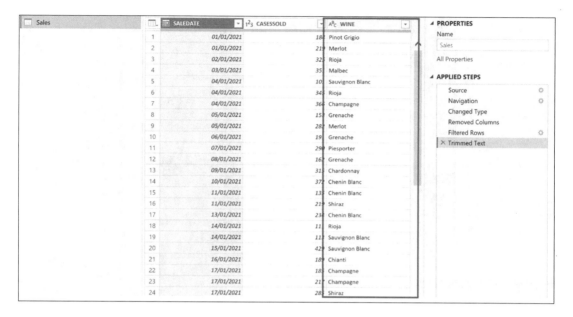

Figure 5-13. *A flat table imported into Power Query comprising the WINE column*

After importing the flat table, we must first perform any filtering transformations that are required. In Figure 5-13, you can see in the APPLIED STEPS pane that we have done this. Note also that we have trimmed the text in the WINE column to remove any trailing spaces which would cause problems when we attempt to relate this field to its counterpart. You will find the "Trim" option on the **Transform** tab, under the **Format** button. We must now duplicate the query by right-clicking the query name in the QUERIES pane and selecting **Duplicate**, as shown in Figure 5-14. Rename your duplicated query.

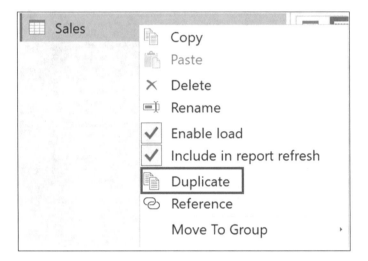

Figure 5-14. *Duplicate the fact table query*

Now you must select the column or columns that will comprise the dimension. In our case, it's only the WINE column, but you would select all the columns to be contained in the dimension. On the **Home** tab, use the **Remove Columns** button to **Remove Other Columns**; see Figure 5-15.

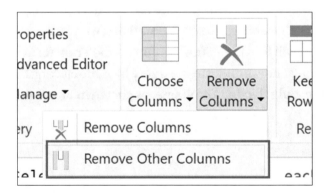

Figure 5-15. *Remove all the columns except the columns that you want to keep*

Dimensions must contain unique values in the linking column (in our example the WINE column), so now on the **Remove Rows** button, select **Remove Duplicates** as shown in Figure 5-16.

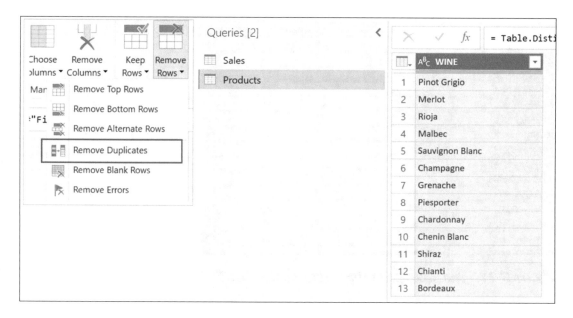

Figure 5-16. *Remove the duplicates to generate the dimension*

Now you have generated your dimension you can relate it to the fact table in a many-to-one relationship. It's worth observing that a dimension can contain just a single column of unique values. Nevertheless, it is still required by the star schema.

However, we must all be pragmatic at times. Sometimes it's just not feasible to generate many dimensions in your model. But what happens when you discover you need a dimension to perform specific calculations but it doesn't exist in your data model? For instance, our Sales fact table may contain the CUSTOMER NAME field because we have no Customers dimension; see Figure 5-17.

SALE DATE	WINESALES NO	SALESPERSON ID	WINE ID	QUANTITY	REVENUE	CUSTOMER NAME
02 January 2018	1	4	2	108	$10,800	Melbourne Ltd
02 January 2018	2	1	12	132	$6,600	Beaverton & Co
03 January 2018	3	5	2	90	$9,000	Miyagi & Co
06 January 2018	5	6	13	291	$22,698	Cape Canaveral Ltd
06 January 2018	4	5	3	216	$16,200	Fremont & Sons
09 January 2018	7	4	7	192	$7,680	Brooklyn & Co
09 January 2018	6	1	8	159	$4,770	El Cajon & Sons
10 January 2018	8	3	2	106	$10,600	Melbourne Ltd
14 January 2018	9	1	7	226	$9,040	Eilenburg Ltd
19 January 2018	10	4	9	227	$8,853	Clifton Ltd
22 January 2018	11	3	11	227	$10,215	Issaquah & Co
26 January 2018	13	4	7	223	$8,920	Canoga Park Ltd
26 January 2018	12	3	8	138	$4,140	Hawthorne Bros
27 January 2018	14	1	10	140	$5,600	East Orange & Co
01 February 2018	15	2	8	116	$3,480	Old Saybrook Ltd

Figure 5-17. *The Sales fact table contains the CUSTOMER NAME field*

But now we must calculate how many sales we have where the total quantity value for any customer is greater than 3,000 and display this value in a card visual. If we first write this measure:

```
Total Qty = SUM ( Sales[QUANTITY])
```

We can now perform this calculation:

```
Total Qty GT 3000 Wrong =
COUNTROWS ( FILTER ( Sales,[Total Qty] > 3000) )
```

However, this will calculate "Total Qty" values greater than 3,000 for *each transaction,* not the totals for each customer. We have no quantity values greater than 3,000 in the QUANTITY column of the Sales fact table. If we had a customer's dimension, we could author this simple measure that calculates the "Total Qty" measure for each row in the Customers dimension:

```
Total Qty GT 3000 Using a Dimension =
COUNTROWS ( FILTER(Customers,[Total Qty]>3000 ) )
```

But we don't have such a dimension, and therefore, we must find a way of grouping each customer's transactions so we can aggregate the total and then find those totals that are greater than 3,000. In the absence of a dimension, we can use the SUMMARIZE function to generate a "dimension" on the fly, as follows:

```
Total Qty GT 3000 =
COUNTROWS (
    FILTER ( SUMMARIZE ( Sales, Sales[CUSTOMER NAME] ),
            [Total Qty] > 3000 ))
```

You can see the results of these measures in Figure 5-18.

CUSTOMER NAME	Total Qty	Total Qty GT 3000 Wrong	Total Qty GT 3000 Using a Dimension	Total Qty GT 3000	
Ballard & Sons	4,542		1	1	
Barstow Ltd	1,934				
Beaverton & Co	2,273				
Black Ltd	2,139				
Black River & Co	3,543		1	1	
Bluffton Bros	3,309		1	1	
Branch Ltd	3,607		1	1	
Brooklyn & Co	2,725				
Brooklyn Ltd	2,011				
Brown & Co	2,346				
Burlington Ltd	3,319		1	1	
Burningsuit Ltd	4,840		1	1	
Busan & Co	3,786		1	1	
Canoga Park Ltd	4,297		1	1	
Cape Canaveral Ltd	2,373				

43

Total Qty GT 3000

Figure 5-18. *In the absence of a Customers dimension, we can use the SUMMARIZE function in the "Total Qty GT 3000" measure*

However, it would be more sensible to generate a Customers dimension and therefore prevent the requirement for more complex DAX expressions.

Q35. How Can I Manage Many-to-Many Relationships?

This question, like the answer to it, is not something we can explain in a few short paragraphs because there are a number of different facets and angles from which we must consider the question.

If you're building a data model in Power BI and creating relationships between tables, you may come upon the message shown in Figure 5-19.

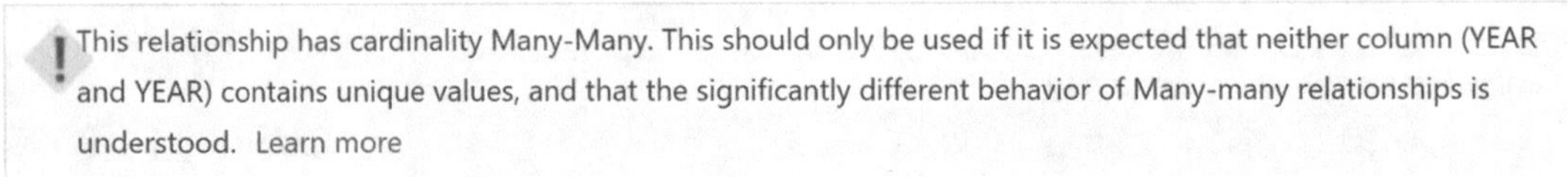

> ! This relationship has cardinality Many-Many. This should only be used if it is expected that neither column (YEAR and YEAR) contains unique values, and that the significantly different behavior of Many-many relationships is understood.   Learn more

Figure 5-19. *This message will display if you attempt to relate columns both of which have duplicate values*

The reason this message has cropped up is that you are attempting to link two columns that both contain duplicate values. But by reading the message you may feel that you're no further on in understanding what exactly you're doing wrong. What, for instance, does "cardinality Many-Many" mean? And what is the "Significantly different behavior" about which they're warning you? Managing many-to-many relationships in your data model can be complicated, but worth understanding if you want to create robust data models in Power BI that stay within the constraints of the star schema.

Let's start with the first complication; in Power BI data modeling, there are two types of many-to-many relationships. The first type, mentioned in the warning message, we can call *cardinal* many-to-many relationships. These occur when you attempt to link two columns both of which hold duplicates. The second type we can call *conventional* many-to-many relationships which arise when you have irregularities in the fact table.

There is a setting in the **Options** area of Power BI Desktop (under **Options and settings** on the **File** tab), whereby relationships are automatically detected, and this is enabled by default. This option is under **CURRENT FILE** and the **Data Load** category; see Figure 5-20.

Figure 5-20. *The automatic detection of relationships option is enabled*

If this is the case, cardinal many-to-many relationships often present themselves automatically when attempting to generate a data model, and the default is to use bi-directional filtering with many-to-many relationships. For this reason, perhaps the biggest problem is that they work. However, just because they are a quick fix, doesn't mean you should use them. The golden rule as set out in Q33, "Why Do I Need a Star Schema" is that data models should comprise star schemas where filters only flow from dimensions to the fact table and never in both directions. As soon as you start implementing cardinal many-to-many relationships with bi-directional filtering, you have immediately broken this rule and violated the integrity of the star schema. You have, in short, lost control of your data model. However, we must find a way to solve the problem of generating relationships where the cardinality of the data prevents us from doing so.

To do this, let's again explore a scenario. For instance, in our Salespeople Targets table, we have recorded the yearly targets for each salesperson; see Figure 5-21.

YEAR	TARGET	SALESPERSON ID
2018	$588,971	4
2018	$378,316	1
2018	$534,524	2
2018	$370,215	6
2018	$522,688	5
2018	$285,354	3
2019	$565,759	2
2019	$403,744	3
2019	$470,983	5
2019	$298,560	1
2019	$595,663	4
2019	$588,753	6

Figure 5-21. *Our salespeoples' targets*

This table is related to the SalesPeople table in a many-to-one relationship using the SALEPERSON ID columns; see Figure 5-22.

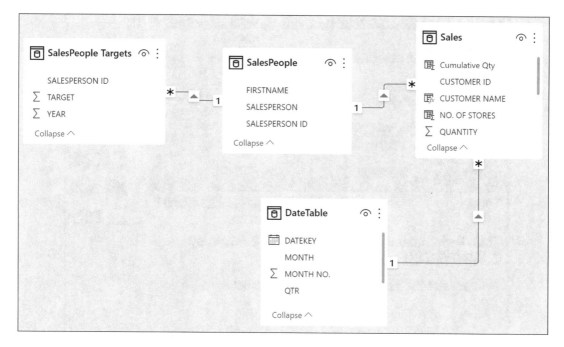

Figure 5-22. *The SalesPeople Targets table is related to the SalesPeople table*

We would like to compare yearly sales and yearly targets for our salespeople and be able to show this in a table visual or column chart, as depicted in Figure 5-23.

YEAR	SALESPERSON	Total Revenue	SalesPeoples Yearly Target ▼
2022	Charron	$542,625	$653,160
2022	Denis	$661,286	$592,675
2022	Blanchet	$258,909	$546,411
2022	Reyer	$633,456	$494,815
2022	Leblanc	$593,300	$467,079
2022	Abel	$466,688	$393,247
Total		**$3,156,264**	**$3,147,387**

YEAR
☐ 2018
☐ 2019
☐ 2020
☐ 2021
■ 2022
☐ 2023

Total Revenue and SalesPeoples Yearly Target by SALESPERSON

● Total Revenue ● SalesPeoples Yearly Target

Figure 5-23. *Table and column chart showing our salespeoples' yearly targets*

The problem will be how to get both the target values and the total revenue value in the same visual against each year. We will get different calculations depending from which table the year value column comes, as shown in Figure 5-24. We get an incorrect "TARGET" value if the year comes from the DateTable, and we get an incorrect "Total Revenue" value if the year comes from the SalesPeople Targets table.

	Year from the DateTable					Year from the SalesPeople Targets Table			
YEAR	SALESPERSON	Total Revenue	TARGET	^	YEAR	SALESPERSON	Total Revenue	TARGET	^
2018	Abel	$218,341	$2,377,172		2018	Abel	$2,082,865	$378,316	
2019	Abel	$266,999	$2,377,172		2019	Abel	$2,082,865	$298,560	
2020	Abel	$314,781	$2,377,172		2020	Abel	$2,082,865	$403,404	
2021	Abel	$625,842	$2,377,172		2021	Abel	$2,082,865	$337,922	
2022	Abel	$466,688	$2,377,172		2022	Abel	$2,082,865	$393,247	
2023	Abel	$190,214	$2,377,172		2023	Abel	$2,082,865	$565,723	
2018	Blanchet	$151,937	$3,303,359		2018	Blanchet	$1,864,409	$534,524	
2019	Blanchet	$271,799	$3,303,359		2019	Blanchet	$1,864,409	$565,759	
2020	Blanchet	$462,061	$3,303,359		2020	Blanchet	$1,864,409	$636,537	
2021	Blanchet	$472,264	$3,303,359		2021	Blanchet	$1,864,409	$608,875	
2022	Blanchet	$258,909	$3,303,359		2022	Blanchet	$1,864,409	$546,411	
Total		**$13,698,595**	**$16,790,507**	˅	**Total**		**$13,698,595**	**$16,790,507**	˅

Figure 5-24. *Incorrect calculations depending on the table that holds the year*

Looking at the model in Figure 5-22, we can see the problem. If we take Year from the Date Table, the Year filter is propagated to the Sales table to show sales for each year, but this filter is not propagated onward to the SalesPeople Targets table via the Salespeople table. If we take Year from the SalesPeople Targets table, this filter won't propagate to any other tables. What must be done is to filter the years in the DateTable that will filter the dates in the Sales table, but will *also* filter the years in the SalesPeople Targets table.

The simplest solution is to implement a cardinal many-to-many relationship by relating the Year field in the SalesPeople Targets table to the Year column in the DateTable; see Figure 5-25.

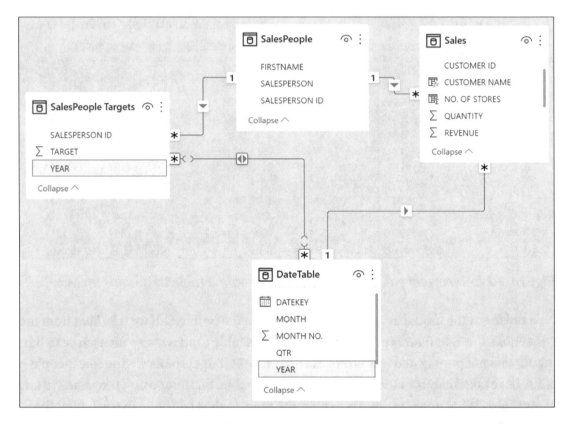

Figure 5-25. *Generating a cardinal many-to-many relationship isn't good practice*

This will result in the warning message shown in Figure 5-19 because in both tables the year values are duplicated, and when generated, it will render an anomaly in the star schema. We must also take note of what else the message tells us; you should only use this relationship if the "significantly different behavior" of many-many relationships is understood. Because they behave differently, cardinal many-to-many relationships are known as "limited" relationships indicated by the notches at either end of the linking line.

Note   There is more information on "limited" relationships, here:

`www.sqlbi.com/articles/strong-and-weak-relationships-in-power-bi/`

It would seem more sensible, therefore, to find an alternative solution that doesn't interfere with the structure of our data model. To do this, we can use the TREATAS function as follows:

```
SalesPeoples Yearly Targets =
CALCULATE ( SUM ('SalesPeople Targets'[TARGET] ),
TREATAS (
VALUES ( DateTable[Year]),'SalesPeople Targets'[YEAR] ) )
```

The "SalesPeoples Yearly Targets" measure creates a "virtual" relationship that matches the year values in the Date Table with the year values in the SalesPeople Targets table.

Notice the VALUES function used as a table expression to create a one-column table (often with only one-row) containing the YEAR value from the DateTable in the current filter context, which is "**2018**" in the first evaluation for "**Abel**." This one-row, one-column table is used to filter the YEAR column in the Salespeople Targets table to equal "**2018**." It's important therefore that we use the YEAR column from the DateTable in the visual.

As an alternative to visualizing the target values in a column chart, you may prefer to show them as horizontal lines; see Figure 5-26. In Q5, "How Do I Show Target Lines on Columns or Bar Charts?" we learned how we could use a line and stacked column chart to generate these lines. However, when generating the secondary Y-axis on which the "SalesPeoples Yearly Target" measure is plotted, you may need to set the minimum value to **0** on the **Range** formatting card. This is so the scales on both Y-axes match.

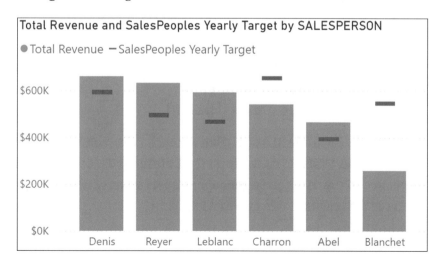

Figure 5-26. *An alternative way to visualize the targets*

If the requirement is to show monthly targets rather than yearly, the "SalesPeople Targets" table might look similar to the values in Figure 5-27.

SALESPERSON ID	YEAR	MONTH	TARGET
1	2018	Jan	$18,722
1	2018	Feb	$13,309
1	2018	Mar	$13,200
1	2018	Apr	$11,830
1	2018	May	$14,429
1	2018	Jun	$16,753
1	2018	Jul	$10,999
1	2018	Aug	$10,771
1	2018	Sep	$13,227
1	2018	Oct	$10,042
1	2018	Nov	$12,928
1	2018	Dec	$17,501
2	2018	Jan	$14,429
2	2018	Feb	$16,753
2	2018	Mar	$10,999

Figure 5-27. *The Salespeople now have monthly targets*

To plot the targets against the monthly sales revenue, this would be the measure:

```
SalesPeoples Monthly Target =
CALCULATE ( SUM ('SalesPeople Targets'[TARGET] ),
    TREATAS (
      SUMMARIZE ( DateTable, DateTable[Year], DateTable[MONTH] ),
              'SalesPeople Targets'[YEAR],
              'SalesPeople Targets'[MONTH] ) )
```

Here we must use the SUMMARIZE function to build a two-column virtual table comprising the YEAR and MONTH values from the DateTable. This virtual table is used to filter the respective columns in the SalesPeople Targets table, which now also contains months. You can see the results of this measure in Figure 5-28.

YEAR	MONTH	SALESPERSON	Total Revenue	SalesPeoples Monthly Target
2023	Jan	Abel	$31,651	$18,722
2023	Feb	Abel	$60,940	$13,309
2023	Mar	Abel	$60,348	$13,200
2023	Apr	Abel	$37,275	$11,830
2023	Jan	Blanchet	$71,829	$17,765
2023	Feb	Blanchet	$49,040	$18,180
2023	Mar	Blanchet	$12,800	$10,414
2023	Apr	Blanchet	$113,770	$14,717
2023	Jan	Charron	$81,996	$15,292
2023	Feb	Charron	$37,095	$13,737
2023	Mar	Charron	$87,278	$19,120
2023	Apr	Charron	$71,009	$17,843
2023	Jan	Denis	$37,643	$16,851
2023	Feb	Denis	$81,800	$12,914
2023	Mar	Denis	$87,266	$10,352

Figure 5-28. *Salespeoples' monthly targets*

By using the TREATAS function in this way, we have solved our many-to-many dilemma *programmatically* and left the star schema intact.

Now let's move on and explore the conventional many-to-many relationship which exposes itself readily in any star schema. All dimensions have a many-to-many relationship between each other. For example, each product has been sold to many customers, and each customer has bought many products. Problems will arise when we need to pass filters from the fact table back up to a dimension. As we know, in a star schema, filters only flow from dimensions to the fact table.

For example, in the Customers dimension, we have the NO. OF STORES column; see Figure 5-6. We would like to calculate the number of stores in which we've sold each wine. We might create this measure:

```
Total Stores =
SUM ( Customers[ NO. OF STORES] )
```

However, as you can see in Figure 5-29, this measure does not work because the filters on the Products dimension only propagate to the fact table and do *not* propagate onward to the Customers dimension.

WINE	Total Stores
Bordeaux	1181
Champagne	1181
Chardonnay	1181
Chenin Blanc	1181
Chianti	1181
Grenache	1181
Lambrusco	1181
Malbec	1181
Merlot	1181
Piesporter	1181

Figure 5-29. *The "Total Stores" measure does not return the correct results*

To resolve this problem, there are three options:

1. Edit the relationship between the Sales table and the Customers table to be bi-directional. You can do this by double-clicking the relationship linking line to open the **Edit Relationship** pane and then, in the "**Cross filter direction**" dropdown, selecting "Both."

2. Use the CROSSFILTER to *programmatically* change the direction of the filter propagation.

3. Use table expansion to *programmatically* pass filters from the sales table onto the Customers dimension.

Let's take each of these alternatives in turn. Making the relationship bi-directional is not a good idea because this interferes with the integrity of the star schema. All we require is a remedy for this particular calculation and not for other calculations that use this relationship.

Using the CROSSFILTER function might be a better solution because it changes the direction of filter propagation only in the execution of the measure in which it is used. This would be the measure we could use and we can see the result in Figure 5-30.

```
Total Stores =
CALCULATE (
    SUM ( Customers[NO. OF STORES] ),
    CROSSFILTER ( Sales[CUSTOMER ID],
  Customers[CUSTOMER ID], BOTH ))
```

WINE	Total Stores
Bordeaux	244
Champagne	777
Chardonnay	748
Chenin Blanc	561
Chianti	590
Grenache	715
Malbec	499
Merlot	581
Piesporter	566
Pinot Grigio	606
Rioja	625
Sauvignon Blanc	660
Shiraz	527
Total	**1,181**

Figure 5-30. *Using CROSSFILTER will programmatically reverse filter propagation and return a result*

However, this measure returns an incorrect value on the Total row. Many of the same customers will have bought each wine, so we know that the total of **1,181** will not be the sum of the values above. However, you might think this value looks about right and so believe it. The value in the Total row should be the total number of stores in which we've sold all our products, but in the Customers table we have five customers to whom we've sold no products. If we "show items with no data" in a table visual where we calculate the "Total Revenue" measure, we can see who they are. See Figure 5-31.

CUSTOMER NAME	Total Revenue
Acme & Sons	
Bloxon Bros.	
Jones Ltd	
Sainsbury's	
Smith & Co	
Cheney & Co	$41,660
Port Hammond Bros	$50,170
Kassel & Sons	$67,439
Charlottesville & Co	$77,190
Rhodes Ltd	$88,086
Victoria Ltd	$93,788
Palo Alto Ltd	$95,801
Morristown & Co	$96,208
Black Ltd	$98,080
Beaverton & Co	$99,240
Warrnambool Ltd	$100,304
Yokohama & Co	$104,395
Total	**$13,698,595**

Figure 5-31. *Some customers have no sales*

The value of **1,181** shown in the Total row includes the stores for these customers. We can see these values in the Customers table in the NO. OF STORES column; see Figure 5-32.

CUSTOMER NAME	REGION ID	NO. OF STORES
Acme & Sons	2000	4
Jones Ltd	100	5
Bloxon Bros.	2000	11
Sainsbury's	100	24
Smith & Co	100	25

Figure 5-32. *Customers with no sales have values in the NO. OF STORES column*

We haven't sold any wine to these customers, so clearly their stores shouldn't be included in the Total number of stores in which we've sold our wines. Our total is out by **69**. The reason the value on the Total row is incorrect is because when the measure arrives at the evaluation of the Total row, the filters are removed from the WINE column of the Products dimension, and therefore, there is no filter to propagate to the Customers dimension. With no filters propagated, it sums all the values in the NO. OF STORES column. In other words, bi-directional filters are only active if filters are active. What we can do here is use the filter on the fact table to filter the Customers table using a concept known as "table expansion." This is the measure that will give you the correct total:

```
Total Stores =
CALCULATE (
    SUM ( Customers[NO. OF STORES] ),Sales )
```

Note   For more information on "table expansion" in DAX, visit my blog, www.burningsuit.co.uk/blog/dax-table-expansion-explained.

We can see in Figure 5-33 that the number of stores in which we have sold our products is now calculated correctly.

WINE	Total Stores
Bordeaux	244
Champagne	777
Chardonnay	748
Chenin Blanc	561
Chianti	590
Grenache	715
Malbec	499
Merlot	581
Piesporter	566
Pinot Grigio	606
Rioja	625
Sauvignon Blanc	660
Shiraz	527
Total	**1,112**

Figure 5-33. *The "Total Stores" measure is now calculated correctly*

We have been exploring a many-to-many relationship that occurs naturally out of the structure of the star schema where dimensions have a many-to-many relationship between them. What we must remember here is that the fact table remains the linking table between all the dimensions. However, there are some scenarios where we must generate a fact table to be used as the link between tables that present themselves as dimensions in many-to-many relationships. Such a fact table is sometimes referred to as a "factless fact table." Consider the following scenario. We have recorded the sales channels for our sales transactions in a "Channel" dimension as shown in Figure 5-34.

"Channel" Dimension	
CHANNEL ID ▼	CHANNEL ▼
1	Phone
2	Internet Search
3	In Person
4	Email
5	Recommendation

Figure 5-34. *Our sales channels*

Each sales transaction can have one or more sales channels associated with it. Therefore, each transaction has many channels, and each channel has many transactions, a conventional many-to-many relationship. The question is, how can we uncover the profitability of selling through each channel?

Initially, the CHANNEL ID and CHANNEL values may have been stored in the Sales fact table which would have rendered duplicated rows when there is more than one channel for each transaction; see Figure 5-35. This would have bloated the row count of the fact table considerably.

WINESALES NO ▼	CHANNEL ID ▼	CHANNEL ▼	ALE DATE ▼	SALESPERSON ID ▼	CUSTOMER ID ▼	WINE ID ▼	QUANTITY ▼
1	1	Phone	01/01/2016	5	41	7	182
2	2	Internet Search	01/01/2016	3	52	6	75
3	2	Internet Search	02/01/2016	3	60	2	146
4	3	In Person	02/01/2016	6	49	4	131
4	5	Recommendation	02/01/2016	6	49	4	131
5	5	Recommendation	02/01/2016	5	51	1	122
6	4	Email	02/01/2016	6	34	11	111
7	1	Phone	02/01/2016	2	79	13	107
7	3	In Person	02/01/2016	2	79	13	107
8	1	Phone	02/01/2016	2	52	10	89
8	2	Internet Search	02/01/2016	2	52	10	89
9	5	Recommendation	03/01/2016	5	47	12	221

Figure 5-35. *The CHANNEL ID field in the fact table renders duplicate rows*

What we can do, therefore, is extract the WINESALES NO and CHANNEL ID columns from the fact table into their own table and use a relationship to the Sales table to associate each transaction with its sales channels. In Figure 5-36, we've named this table "Channel Bridge." This will be our factless fact table.

"Channel Bridge" Fact Table	
WINESALES NO	CHANNEL ID
1	1
2	2
3	2
4	3
4	5
5	5
6	4
7	1
7	3

Figure 5-36. *The WINESALES NO and CHANNEL ID fields have been extracted into their own fact table, "Channel Bridge"*

We can now relate the Channel Bridge table to the Channel dimension (Figure 5-35) and to the Sales "dimension," both using many-to-one relationships; see Figure 5-37.

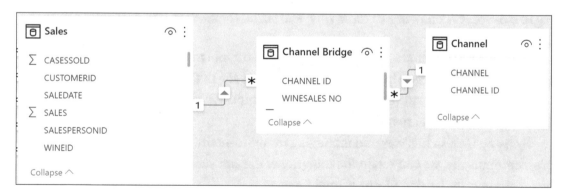

Figure 5-37. *Building in the "Channel Bridge" factless fact table*

Because the Sales table sits on the one side of the relationship and the Channel Bridge table sits on the many, the Channel Bridge has become a factless fact table, while the Sales table now takes on the role of a dimension.

To calculate the sales revenue for each channel, we can now either use the CROSSFILTER function, or we can pass the filter onto the Channel Bridge table using table expansion:

```
Sales Value =
CALCULATE (
    [Total Revenue],
    CROSSFILTER ( Sales[WINESALES NO],
      'Channel Bridge'[WINESALES NO], BOTH ))
```

```
Channel Sales Value #2=
CALCULATE ( Total Revenue], 'Channel Bridge')
```

Either of these measures would return the same result; see Figure 5-38.

CHANNEL	Sales Value
Email	$9,487,830
In Person	$9,212,044
Internet Search	$9,021,528
Phone	$9,587,318
Recommendation	$9,071,165
Total	**$13,698,595**

Figure 5-38. *Calculating the sale revenue value for each sales channel*

What we can conclude from exploring a conventional many-to-many relationship is that they can be managed by generating a table that sits in the middle, whereby we can relate the two tables using many-to-one relationships. We can then use DAX measures to perform calculations across the tables.

By using DAX in this way and managing many-to-many relationships programmatically, we can retain the integrity of the star schema, and this is always a better strategy than using bi-directional filtering.

Q36. Why Do I Get Dotted Lines Between Tables When I Create a Relationship?

This is because one thing that Power BI prevents is *ambiguity* in the data model. This is where there would be multiple paths through which filters could propagate. Therefore, if you attempt to relate a dimension to two or more other dimensions, this will result in an *inactive* relationship being created, indicated by a dotted line. For example, in Figure 5-39, we have related the SalesPeople dimension to both the Customers dimension *and* the Regions dimension, and this results in an inactive relationship between SalesPeople and Regions.

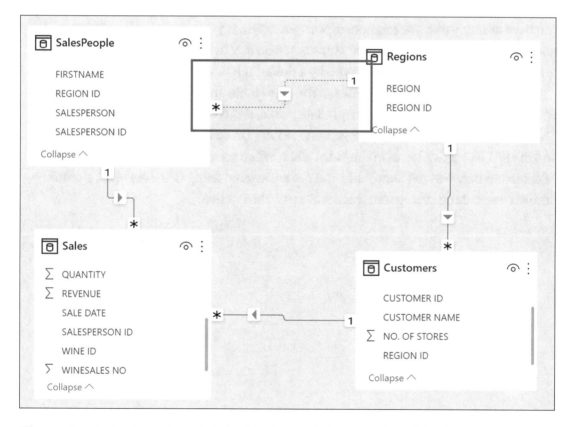

Figure 5-39. *An inactive relationship is created to avoid ambiguity*

You can see the problem and why we must have one relationship inactive. If both relationships were active, if we filter by REGION, the filter will propagate to the Sales table via the SalesPeople dimension, or it will propagate via the Customers dimension. However, it can't do both. We must decide which route the filter will take because we can activate the inactive relationships using DAX. For filters to propagate via the SalesPeople dimension, we must use the USERELATIONSHIP function as follows:

```
Total Revenue for SalesPeople Regions =
CALCULATE (
    [Total Revenue],
    USERELATIONSHIP ( SalesPeople[REGION ID], Regions[REGION ID] ) )
```

The USERELATIONSHIP function activates inactivate relationships by referencing the columns to be linked, from the many side of the relationship to the one side. It must always be nested inside CALCULATE.

It's only possible to have *one active* relationship between any two tables, but you can have as many *inactive* relationships as you want. If you attempt to build a second relationship or subsequent relationships between any two tables, all but the first relationship will be inactive, indicated by a dotted relationship line.

Consider the relationships between the Sales table and DateTable in Figure 5-40. We now have two date columns in our Sales table, SALE DATE and ORDER DATE. The first relationship was established between the DATE KEY column in the DateTable and the SALE DATE column in the Sales table. When we attempted to create a second relationship between the DateTable and the Sales table using ORDER DATE, a dotted line was created indicating that this relationship is inactive.

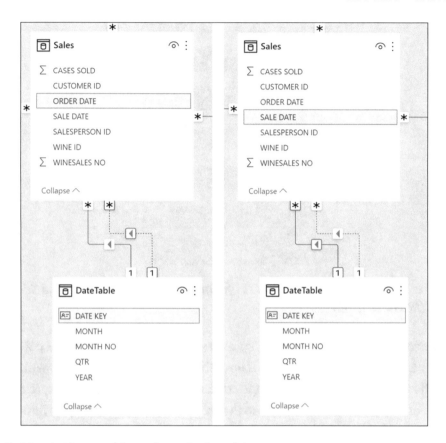

Figure 5-40. *Active and inactive relationships*

All measures will use the active relationship by default, but we have already discovered that we can use the USERELATIONSHIP function to activate inactive relationships. For example, if we build a table visual containing the YEAR and MONTH columns from the DateTable (Figure 5-41), we can find the number of sales in each month using this measure:

```
No. of Sales =
COUNTROWS ( Sales )
```

YEAR	MONTH	No. of Sales
2021	Feb	51
2021	Mar	58
2021	Apr	55
2021	May	55
2021	Jun	61
2021	Jul	55
2021	Aug	46
2021	Sep	51
2021	Oct	65
2021	Nov	43
Total		**2,197**

Figure 5-41. *Using YEAR and MONTH from the DateTable filters the SALE DATE column in the Sales table*

In this visual, the "No. of Sales" measure filters the YEAR and MONTH columns in the DateTable, which propagates to the Sales table using the *active* relationship and therefore filters the SALE DATE column for that year and month. However, to calculate the number of orders, we will need to use the *inactive* relationship so that filters propagate from the DateTable to the ORDER DATE column:

```
No. of Orders =
CALCULATE (
    COUNTROWS ( Sales ),
    USERELATIONSHIP ( Sales[ORDER DATE], DateTable[DATE KEY] ))
```

When this measure is evaluated, the year and month filtered in the DateTable is propagated to the Sales table to cross-filter the ORDER DATE column to find the orders in each month as depicted in Figure 5-42.

YEAR	MONTH	No. of Sales	No. of Orders
2021	Feb	51	49
2021	Mar	58	66
2021	Apr	55	42
2021	May	55	59
2021	Jun	61	57
2021	Jul	55	60
2021	Aug	46	46
2021	Sep	51	54
2021	Oct	65	56
2021	Nov	43	49
2021	Dec	39	28
Total		**2,197**	**2,197**

Figure 5-42. *Calculating the number of sales and number of orders*

The dotted relationship lines in Model view indicate an inactive relationship that Power BI will generate on your behalf to prevent ambiguity in the way filters propagate through the data model.

Q37. Why Does Filtering the Fact Table Also Filter a Dimension?

Well, actually this would never happen. As we are continually repeating, one of the major factors that underpin a good Power BI data model is when filters only propagate from the one side of a relationship to the many, that is, from dimension tables to the fact table. This is known as single-directional filtering. However, even when such a model has been designed, it may appear that filters are, in fact, passing from the fact table back up to a dimension. It may appear that when you filter the fact table, a dimension will respond to this filter as if the filter has propagated from the many side of the relationship to the one side.

Let's take an example of this behavior. Our Sales fact table is related to the Customers dimension via the CUSTOMER ID columns in both tables; see Figure 5-43.

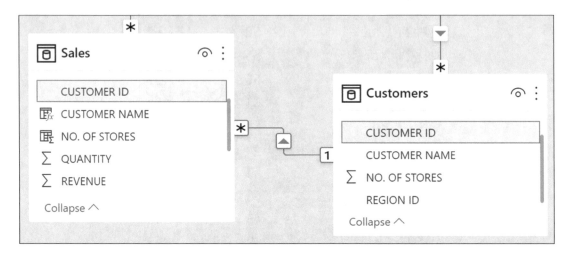

Figure 5-43. *The Customers dimension is related to the Sales fact table in a many-to-one relationship*

We have a measure that sums the QUANTITY column in the Sales table:

```
Total Qty = SUM ( Sales[QUANTITY] )
```

In Figure 5-44, we have created a table visual that holds the CUSTOMER ID column from the Sales fact table and the "Total Qty" measure. We have then created a second table visual that holds the CUSTOMER ID column from the Customers table, again with the "Total Qty" measure. A slicer is then populated with the CUSTOMER ID column from the *Sales* table.

CUSTOMER ID from Sales Table

CUSTOMER ID from Sales Table   CUSTOMER ID from Customers Table

CUSTOMER ID		CUSTOMER ID	Total Qty	^	CUSTOMER ID	Total Qty	^
☐ 1		1	3,938		1	3,938	
☐ 2		2	3,065		2	3,065	
☐ 3		3	2,373		3	2,373	
☐ 4		4	2,139		4	2,139	
		5	3,668		5	3,668	
☐ 5		6	3,367		6	3,367	
☐ 6		7	2,771		7	2,771	
☐ 7		8	1,258		8	1,258	
		9	2,346		9	2,346	
☐ 8		10	3,942		10	3,942	
☐ 9		11	3,639		11	3,639	
☐ 10		12	3,110		12	3,110	
		13	4,264		13	4,264	
☐ 11		14	2,570		14	2,570	
☐ 12		15	3,384		15	3,384	
☐ ₁₂		**Total**	**251,910**	^v	**Total**	**251,910**	^v

Figure 5-44. *The CUSTOMER ID field comes from the Sales table and from the Customers table*

If we now use the slicer to filter CUSTOMER ID **4**, the CUSTOMER ID column from the Sales fact table is filtered accordingly, which we would expect, but the slicer also appears to filter the CUSTOMER ID field from the Customers dimension which we would not expect because filters don't pass from the many side of a relationship to the one side as illustrated in Figure 5-45.

CUSTOMER ID from Sales Table   CUSTOMER ID from Customers Table

CUSTOMER ID		CUSTOMER ID	Total Qty		CUSTOMER ID	Total Qty
☐ 1			4	2,139		4	2,139
☐ 2		**Total**	**2,139**		**Total**	**2,139**
☐ 3						
■ 4						
☐ 5						
☐ 6						
☐ 7						
☐ 8						

Figure 5-45. *It appears that the CUSTOMER ID field from the Customers dimension has been filtered by the Sales table*

However, what we are looking at here is simply an illusion. The Customers dimension is not being filtered at all. Because the fact table has been filtered to sales belonging to CUSTOMER ID **4** only, there are no values for the "Total Qty" measure for other customers. By default items with no values don't show in visuals. If we right-click the CUSTOMER ID field in the Columns bucket of the visual, we can turn on "**Show items with no data**" and see that this is true (Figure 5-46).

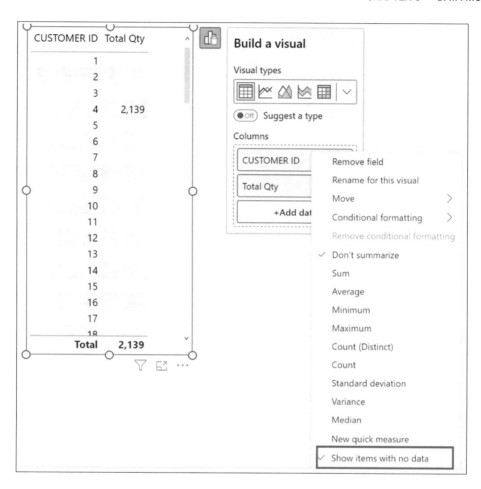

Figure 5-46. *If we "Show items with no data," we can see that there are no values to show for customers not being filtered*

Therefore, we can see that the filter from the Sales fact table is not being propagated to the dimension. It's just because where there are no calculations to show in the table visual, those items are not displayed by default. You can see in Figure 5-47 that there is a difference if we have a slicer that uses the CUSTOMER ID field from the Customers dimension. Even with "Show items with no data" still turned on, when we filter CUSTOMER ID "**4**" from this slicer, this filter is propagated to the fact table, and so both visuals filter accordingly.

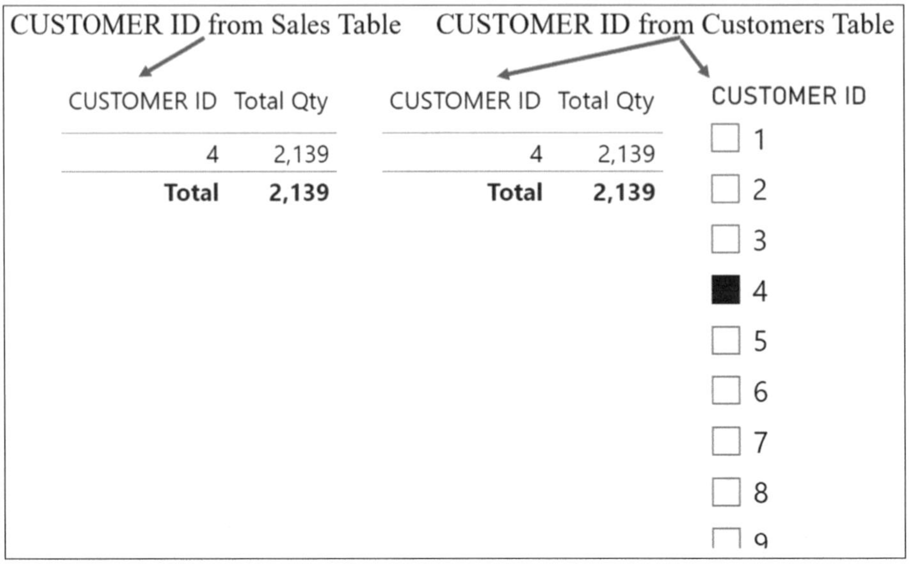

Figure 5-47. *A filter on the Customers table filters both Customers and Sales tables*

However, let's suppose you want to ensure that filtering the Sales fact table directly did not appear to filter the Customers dimension. In that case, there must always be values to show in the Customers dimension. Therefore this measure would do that job:

```
Total Qty #2 =
CALCULATE (
    [Total Qty],
    ALL ( Sales[CUSTOMER ID] ),
    VALUES ( Customers[CUSTOMER ID] )
)
```

CUSTOMER ID from Sales Table   CUSTOMER ID from Customers Table

CUSTOMER ID		CUSTOMER ID	Total Qty		CUSTOMER ID	Total Qty #2
☐ 1		4	2,139		1	3,938
☐ 2		**Total**	**2,139**		2	3,065
☐ 3					3	2,373
■ 4					4	2,139
☐ 5					5	3,668
☐ 6					6	3,367
☐ 7					7	2,771
☐ 8					8	1,258
☐ 9					9	2,346
☐ 10					10	3,942
☐ 11					11	3,639
					12	3,110
					13	4,264
					14	2,570

Figure 5-48. *We can use a DAX measure to always return values from the Customers table when filtering the Sales table*

When we put this measure into the table using the CUSTOMER ID field from the Customers table, we get values for all customers, and the slicer only filters the Sales table (Figure 5-48).

Unless you are using bi-directional filtering (not to be recommended), filters will only propagate from dimension tables to fact tables. However, always keep in mind that by default blank values are never displayed in visuals unless you turn on "show items with no data," and this may give the illusion of a visual being filtered.

CHAPTER 6

DAX Conundrums

With the ever-increasing adoption of Power BI as the preferred data analytics platform, the ability to use DAX is fast becoming a necessary requirement to find and share the important insights into your data, and it's interesting to note that most answers to questions that arise on the Power BI community involve writing DAX code. To find answers, most people resort to searching the Internet, probably because writing the correct DAX continues to be an elusive goal for many Power BI users. Within this chapter, we aim to shed light on the responses to the most frequently asked questions that have come from people who are first embarking on their journey through DAX.

Q38. Why Do I Find DAX Difficult?

It's most likely that your lack of understanding of the fundamental *concepts* of DAX is the reason for this.

Most people first attempt to learn DAX as they learned Excel. They think they can just learn up a few DAX functions, perhaps even nesting one function inside another. They also think they can just copy and paste DAX expressions anywhere in the hope that they will work. Unfortunately, it's not quite as simple as this as they soon find out. It's not long before they realize that what would be a simple calculation in Excel appears impossible to solve using DAX.

One of the first hurdles you'll meet when first introduced to DAX is to understand the use of three different types of expression: calculated columns, measures, and calculated tables.

© Alison Box 2023
A. Box, *A Power BI Compendium*, https://doi.org/10.1007/978-1-4842-9765-0_6

Note With a robust data model, there should be no requirement for generating calculated tables using DAX. The requirement for doing so only arose because, in the early days of DAX, when it resided in Excel's PowerPivot, there was no ETL tool such as Power Query. Therefore, there was a requirement to use DAX for building additional tables and querying data.

Unfortunately, the expressions are not interchangeable. An expression that works in the context of a calculated column will rarely work when put into a measure and the code for a measure will never work if used in a calculated table. Paradoxically, and just to make life complicated, expressions used in measures can also be used in calculated columns. We examine the difference between calculated columns and measures in Q40, "What Is the Difference Between a Calculated Column and a Measure." However, for the purposes of answering the question of why is DAX difficult, most people assume we are referring to the creation of measures, which are considerably more challenging to author than calculated columns.

The problem with DAX, particularly in understanding the DAX measure, is that it's all about *understanding the theory*. For example, take this very short DAX measure:

```
Max of Totals =
MAXX ( Customers, [Total Revenue] )
```

It contains a function, a reference to a table, and a reference to another measure. What could be simpler? However, to appreciate how this expression works, you'll need a good understanding of these DAX concepts: data modeling, filter context, row context, iterators, and context transition. With DAX it's all in the details of the expression. A simple expression can hide a wealth of complexities because, when using DAX it's all about understanding what's going on behind the scenes. Therefore to answer the question, why is DAX difficult, let's explore some of the theories that you must master before you will find DAX in any way easy.

A crucial characteristic of DAX is the inseparable connection between expressions and the structure of the data model. This is something that we explored in detail in Q33, "Why Do I Need a Start Schema?" Here we learned that you must have a clearly defined fact table and dimensions in your data model. The preceding expression, for example, depends on the existence of a Customers dimension. The better structured the model, the easier the DAX calculations. The more you diverge from the star schema, the more complex your DAX expression must be to compensate for anomalies in the schema. Because DAX and the structure of the data model are symbiotic, it also means that a DAX formula that works in one data model may not work in another.

The next milestone in learning DAX is understanding that the values returned by a DAX expression depend on the context in which the expression is being evaluated, which can change as the data is browsed. All DAX measures are evaluated in a *filter context* whereby on the evaluation, the DAX measure filters tables in the data model depending on the construct of the visualization in which the calculation is performed. Therefore, although we can design "global" measures that work across all visualizations, often expressions are designed to work only with a specific visual. We deep-dive into the concept of the filter context in the next question, Q39, "Why Doesn't My Measure Work?"

However, there is another context in which expressions are evaluated. Calculated columns and some measures use the *row context*. Expressions evaluated in the row context reference the values sitting in the current row in the iteration of a table. Measures create a row context if the expression uses an iterating function, such as SUMX or FILTER.

It would be nice to think at this stage that understanding evaluation context, both row context and filter context, would be enough. Unfortunately, DAX will now throw another spanner in the works when you start generating your own expressions. This is because some measures that use the row context will then transition the evaluation into a filter context. This is known as *context transition*. It's beyond the scope of this book to elaborate any further on this challenging concept, but you can check out my blog for further information, `www.burningsuit.co.uk/blog/context-transition-where-row-context-becomes-filter-context`.

If you ask any competent user what their epiphany moment was when learning DAX, they will probably tell you it was the realization that DAX is all about generating *virtual tables* (sometimes called temporary or in-memory tables). It's true that many expressions will reference "real" tables in the data model, but the true capabilities of DAX come to the fore when re-generating these tables in memory. This is done to construct subsets or to control the grouping of data. To build virtual tables, we use a set of DAX functions called *table functions*. These are nested inside other functions that accept a "table" in their parameters. However, the complexity surrounding table functions is that the tables they generate are used for two distinct purposes. If a table function is used inside a function other than CALCULATE, it creates subsets of the original table. However, table functions as filter parameters to CALCULATE generate in-memory tables that are used to filter the data model.

This brings us finally to the last concept that must be tackled to fully understand how DAX works, the final piece in the jigsaw if you like. This is the concept of *table expansion* that explains how filter propagation operates. Most people think that when they create many-to-one relationships between tables, this instigates a "lookup" between what we may refer to as "primary" and "foreign" keys. This isn't how filters perform in memory. The filter context works because on the evaluation of a measure, in memory all the tables that sit on the one side of a relationship are merged into the table that sits on the many, generating one big flat table, or *an expanded table*. It's this expanded table that is filtered by applying filters to the columns from dimensions that now comprise it. Relationships between tables only exist to generate expanded tables. This, again, is why the star schema is so important because table expansion should be only applicable on the fact table. This all may sound a little theoretical. Why is knowledge of table expansion important when working with DAX? One of the great benefits of knowing about table expansion is that you can use this knowledge to control filter propagation through the data model.

Note   For more information on table expansion in DAX, visit my blog, `www.burningsuit.co.uk/blog/dax-table-expansion-explained`.

If you want to get to grips with DAX, all these concepts are mandatory knowledge. Only when you truly understand these core ideas on which the DAX language is built will you ever be able to master the language. There is no shortcut to gaining this knowledge.

Q39. Why Doesn't My Measure Work?

Do your DAX measures return interesting values but not the numbers you want? If the answer is yes, and it happens to all of us, then the trouble likely stems from your failure to accurately identify the filter context in which the measure has been evaluated.

It's essential to understand that all DAX measures are evaluated within a specific filter context. What this means is that when a measure is calculated, filters are applied in memory to tables in the data model. Perhaps then, we should first take a look at the structure of the data model we will use in the following examples, which, as you can see in Figure 6-1, is a star schema with a snowflake dimension which is the Regions table.

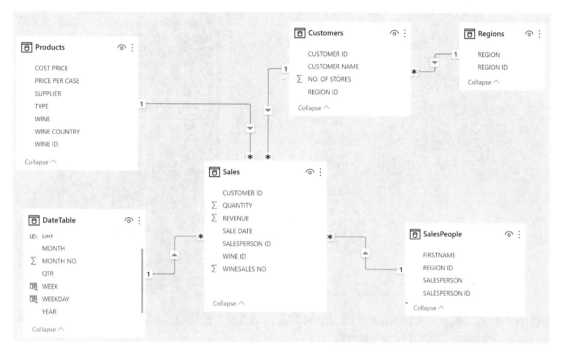

Figure 6-1. *The data model we are using is a star schema and snowflake dimension*

On the evaluation of a measure, filters are generated on this data model depending on three factors that impact the behavior of the visual where the measure is being calculated. These are as follows:

1. Which column or columns are used to categorize the data in the visual?

2. Which columns used in slicers filter the data in the visual?

3. Are there any filters present in the Filters pane that filter the data?

When any of these three elements converge, they create the "filter context" that determines how the measure is evaluated. For example, the "Total Qty" measure sums the QUANTITY column in the Sales table:

```
Total Qty =
SUM ( Sales[QUANTITY] )
```

Consider Figure 6-2, where we are evaluating this measure in a table visual.

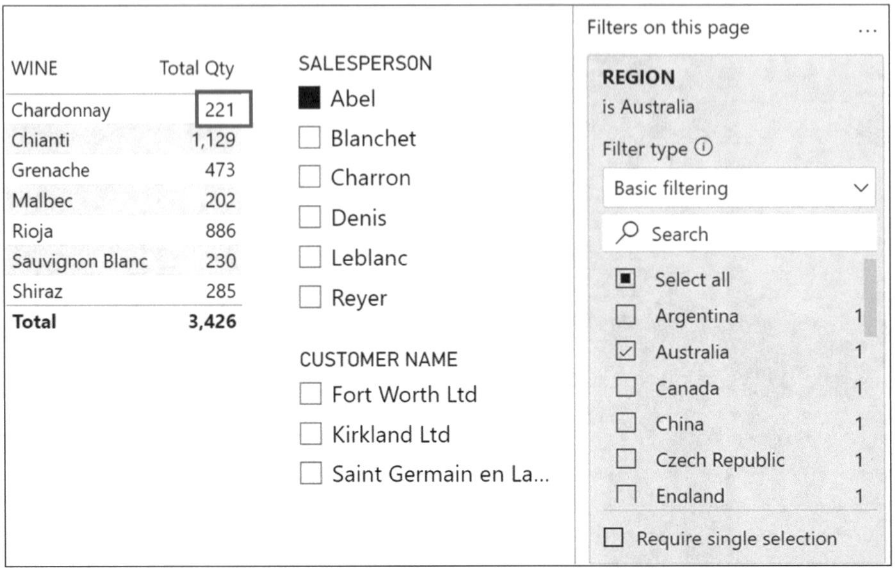

Figure 6-2. *Filters are placed in the WINE, SALESPERSON, and REGION columns to calculate the "Total Qty" value*

To arrive at the value "221" for "Chardonnay," the filter context is generated as follows:

1. The data is being categorized by the WINE column from the Products table, filtering "Chardonnay."

2. In the slicer, the SALESPERSON column in the Salespeople table is filtered for "Abel."

3. In the Filters pane, the REGION column in the Regions table is filtered for "Australia."

The problem is that these filters are hidden from us. If we take a peek at our data model in Model view, we won't be able to see the rows of the tables being reduced as a measure is calculated. Instead, we just have to imagine how they would present themselves in memory, and this would be as shown in Figure 6-3.

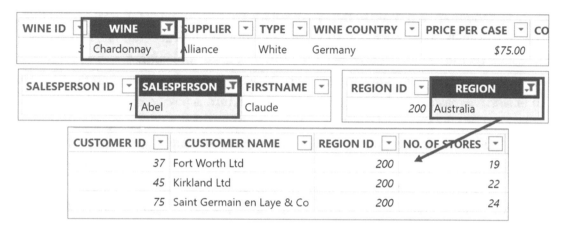

Figure 6-3. *The filters affecting the evaluation of the "Total Qty" measure*

Let's now consider how the structure of the data model allows these filters to propagate to the fact table to sum the values in the QUANTITY column. If you consult Figure 6-4, you will see that the arrow in the linking line of the relationship indicates the direction of this propagation.

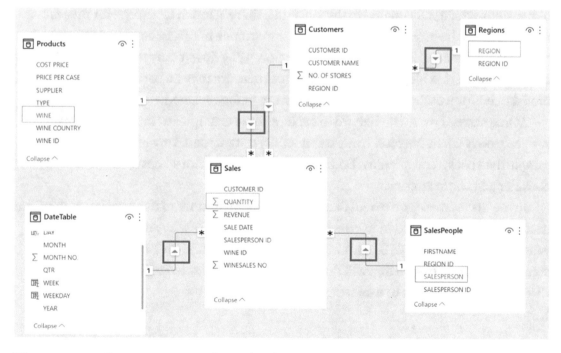

Figure 6-4. *Filters propagate from the dimensions into the fact table*

If "Chardonnay" is filtered in the WINE column of the Products dimension, this filter is passed through to the fact table whose rows are now filtered accordingly. We can see that a filter on the SALESPERSON column from the SalesPeople dimension will similarly reduce the rows to be evaluated in the fact table. Note how a filter placed on the Regions dimension will pass to the Customers table which in turn is passed onto the fact table. Evidence of this is shown in Figure 6-2, where you'll note that, in the slicer containing the CUSTOMER NAME column, values are filtered to show only customers in "Australia."

When the "Total Qty" measure is evaluated for the Total row of the table visual (returning "**3,426**" in Figure 6-2), it does not sum all the values in the rows above for each wine. Instead, the measure is evaluated in a different filter context where all the filters have been removed from the WINE column in the Products table. If there are no filters on the WINE column, the measure will calculate the total quantity for all wines.

The filter context in which the measure is evaluated depends on both how the visual is constructed and the filters that interact with the visual, and this will determine the outcome of the measure. How filters propagate from dimensions to the fact table depends on the shape of the data model, and therefore, the evaluation of the measure and this structure are interdependent. If you are working with a star schema, where dimensions are directly related to the fact table, there is much less scope for measures not to work. We have already alluded to the importance of this issue in Q33 "Why Do I Need a Star Schema?" Returning to our question "Why doesn't my measure work?" you can understand that although syntactically the measure may be correct, it's the data model that is preventing the right numbers from being returned.

At this stage, it would be helpful to explore some examples of intentions going seriously awry when the filter context is not fully understood. However, because our sample data uses a star schema, finding such examples is not so easy. But we can consider just two scenarios.

In the first of these, we would like to calculate the number of different customers that each salesperson has sold to in selected years. We can do this by counting the distinct values in the CUSTOMER ID column using the DISTINCTCOUNT function. When constructing the measure that will do this job, we are presented with a choice of two CUSTOMER ID columns to use; see Figure 6-5.

```
1 Distinct Wines = DISTINCTCOUNT(CUSTOMER ID)
```
```
⊞  Customers[CUSTOMER ID]
⊞  Sales[CUSTOMER ID]
```

Figure 6-5. *Which column do we use in the DISTINCTCOUNT function?*

If we use the CUSTOMER ID column from the Customers table, although a valid expression, this returns the number of distinct customers in the Customers dimension (see Figure 6-6).

YEAR	SALESPERSON	Distinct Customers
☐ 2018		
☐ 2019	Abel	89
	Blanchet	89
☐ 2020	Charron	89
☐ 2021	Denis	89
■ 2022	Leblanc	89
	Reyer	89
☐ 2023	**Total**	**89**

Figure 6-6. *Using CUSTOMER ID from the Customer table is not correct*

The results of this measure may seem obvious to you, but can you explain why in Figure 6-6 we see "**89**" against each Salesperson's name? On the evaluation of each salesperson, there is no filter on the Customers dimension, so the expression will always return the distinct count of all the CUSTOMER ID's in the Customers table. The correct expression is, of course:

```
Distinct Customers = DISTINCTCOUNT ( Sales[CUSTOMER ID] )
```

This measure works, as we can see in Figure 6-7, because the table visual is categorizing the data by the SALESPERSON column, and therefore each salesperson will be filtered accordingly, the filter being propagated to the Sales table. Therefore, by passing the distinct count across the CUSTOMER ID in the Sales table, we can find out the number of different customers to which each salesperson has sold.

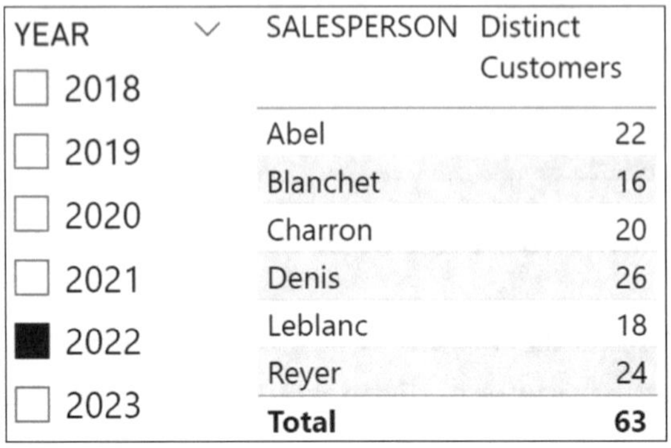

YEAR ∨	SALESPERSON	Distinct Customers
☐ 2018		
☐ 2019	Abel	22
	Blanchet	16
☐ 2020	Charron	20
☐ 2021	Denis	26
■ 2022	Leblanc	18
	Reyer	24
☐ 2023	**Total**	**63**

Figure 6-7. *Using CUSTOMER ID from the Sales table returns the correct values*

In our second scenario, we will consider the problem associated with calculating the average price that our customers paid for their wines. The PRICE PER CASE column resides in the Products dimension, so we might simply find the average of this value:

```
Avg Price = AVERAGE(Products[PRICE PER CASE])
```

However, this measure won't work (Figure 6-8).

CUSTOMER NAME	Avg Price
Acme & Sons	$52.64
Ballard & Sons	$52.64
Barstow Ltd	$52.64
Beaverton & Co	$52.64
Black Ltd	$52.64
Black River & Co	$52.64
Bloxon Bros.	$52.64
Bluffton Bros	$52.64
Branch Ltd	$52.64
Brooklyn & Co	$52.64
Brooklyn Ltd	$52.64
Brown & Co	$52.64

Figure 6-8. *Calculating the average price using the PRICE PER CASE column from the Products dimension*

On the evaluation of each customer, there is no filter on the Products dimension, and therefore the measure finds the average of all the prices in the PRICE PER CASE column. To find the average price each customer paid, we must associate the price of each wine with each customer's transactions. Therefore, we must find a way of populating the Sales table with this price and then pass the average across these values. In Q43 "How Can I Include Values from Dimensions in the Fact Table?" we look more closely at this problem. Suffice to say at this stage this is the correct expression:

```
Avg Price =
AVERAGEX ( Sales, RELATED ( Products[PRICE PER CASE] ) )
```

In the "Avg Price" measure, the expression uses the Sales table that will be cross-filtered to reflect each customer's transactions. The RELATED function pulls the price of each wine through from the Products dimension into the Sales table, and AVERAGEX calculates the average of these values. This is depicted in Figure 6-9.

CUSTOMER NAME	Avg Price
Ballard & Sons	$58.24
Barstow Ltd	$69.00
Beaverton & Co	$44.88
Black Ltd	$45.00
Black River & Co	$42.42
Bluffton Bros	$49.40
Branch Ltd	$69.33
Brooklyn & Co	$41.50
Brooklyn Ltd	$56.89
Brown & Co	$57.67
Burlington Ltd	$52.77

Figure 6-9. *Calculating the average price by associating the PRICE PER CASE with each transaction in the Sales table*

With DAX measures, having a well-structured data model and identifying the filter context in which they will be evaluated is the secret to always ensuring they will work.

Q40. What Is the Difference Between a Calculated Column and a Measure?

Most people would answer this question by stating that a calculated column is what you create in Table view to generate an additional column in a table, whereas a measure is what you put into a visualization to generate a value for reporting. However, it's not quite as simple as that, is it? Because you can create a calculated column that is then used as an implicit measure when the column is put into the values bucket of a visual. So the calculated column has now turned into a measure. You may also have discovered that you can put a measure into a calculated column. We need a better explanation of what this difference is, so let's now see if we can find one.

Calculated columns in DAX are easily understood, especially those familiar with Excel and Excel tables because they resemble "copying down" on an Excel formula. This is because the DAX expression in a calculated column is evaluated for each row in the table, and the values used are restrained to the values of the current row. This evaluation is known as *row context*, and as a result, calculated columns are considered row-level calculations.

In Q39, "Why Doesn't My Measure Work?" we explored the filter context and understood that all measures are evaluated in a specific *filter context*. However, there are two other defining attributes of measures. Firstly, measures are used by all visuals and are always placed in the values bucket. You can rarely construct a visual without an associated measure. Even if you haven't designed your own measures, when you drag and drop a numeric column into the value bucket of a visual, it's converted to an "implicit measure." Secondly, because visuals must aggregate data, all measures return scalar (single) values, typically an aggregated value. Therefore, whereas calculated columns are row-level calculations, measures are *report-level* calculations.

Evaluation context lies at the core of understanding the distinction between a calculated column and a measure. The crucial difference is that calculated columns are evaluated within the row context, whereas measures are evaluated within the filter context. This fundamental distinction explains why these two types of DAX expressions cannot be used interchangeably. Typically, the same DAX expression utilized in a calculated column cannot be directly applied to a DAX measure.

Q41. How Can I Calculate the Average of the Totals?

This is one of those calculations that should be simple to achieve. In Excel, you can easily find the averages of values, whether they are totals or otherwise. However, it's not quite so straightforward when using DAX. Therefore, to answer this question, let's start with first principles and move forward from this.

When learning DAX, we understand early on how to calculate simple averages:

```
Avg Qty = Average ( Sales[QUANTITY] )
```

We know that this calculation is the average quantity for transactions in the Sales table; see Figure 6-10.

SALESPERSON	Total Qty	Avg Qty
Abel	39,384	255.74
Blanchet	36,350	248.97
Charron	38,530	253.49
Denis	49,744	260.44
Leblanc	42,111	268.22
Reyer	45,791	255.82
Total	**251,910**	**257.31**

Figure 6-10. *The "Avg Qty" measure is the average quantity for transactions*

However, there is often a different average that we want to find. We may be using, for example, the "Total Qty" measure, and we want to calculate the average of these totals, which, for our salespeople, is **41,985**; see Figure 6-11.

SALESPERSON	Total Qty
Abel	39,384
Blanchet	36,350
Charron	38,530
Denis	49,744
Leblanc	42,111
Reyer	45,791
Total	**251,910**

We want to find the average of these totals:

41,985

Avg of Totals

Figure 6-11. *We want to calculate the average of the "Total Qty" values*

To generate this and similar calculations, it's vital to have well-defined dimension tables in your data model. This is because we must perform in-memory aggregations (such as defined by the "Total Qty" measure) inside the rows of these dimensions.

To explore this idea, we can simulate the behavior of the measure we are going to design presently by creating this calculated column in the SalesPeople's table:

```
Total Qty Column = [Total Qty]
```

You will notice in Figure 6-12 that it'll calculate the total quantity for each salesperson.

SALESPERSON ID	SALESPERSON	FIRSTNAME	REGION ID	Total Qty Column
1	Abel	Claude	100	39,384
2	Blanchet	Janine	100	36,350
3	Leblanc	Marie	100	42,111
4	Reyer	Maurice	400	45,791
5	Charron	Phillippe	500	38,530
6	Denis	Sally	600	49,744

Figure 6-12. *The "Total Qty" measure in a calculated column*

What you can see is the DAX measure, used in the context of a calculated column, placing a filter on the Sales fact table for each row in the SalesPeople dimension, because all measures are evaluated in a filter context. There are no exceptions to this rule even if the measure is being evaluated in a calculated column.

Note   What I'm briefly describing here is part of a more challenging concept in DAX called *context transition*. For a full explanation of context transition, visit my blog `www.burningsuit.co.uk/blog/context-transition-where-row-context-becomes-filter-context`.

We have placed this measure into a calculated column, but if we place it inside another measure that iterates through the SalesPeople dimension, we will achieve the same result, and, by doing so, we can determine the average of these values.

This would be such a measure:

```
Avg of Totals = AVERAGEX ( SalesPeople, [Total Qty] )
```

Here, each row in the SalesPeople dimension is iterated in memory by AVERAGEX to calculate the "Total Qty" *measure*, just like the calculated column in Figure 6-11. It then finds the average of these values as shown in Figure 6-13.

Note   If you don't have a dimension over which to pass a similar measure, you must generate one programmatically. Typically this is done using the SUMMARIZE function; see Q34, "How Can I Create Dimensions from a Flat Table?"

SALESPERSON	Total Qty	Avg of Totals
Abel	39,384	39,384
Blanchet	36,350	36,350
Charron	38,530	38,530
Denis	49,744	49,744
Leblanc	42,111	42,111
Reyer	45,791	45,791
Total	**251,910**	**41,985**

41,985

Avg of Totals

Figure 6-13.   *The "Average of Totals" measure in a table and card visual*

Because the average of the totals is a single value, it's best visualized in a card visual. If you place the measure into a table visual alongside the salespeople's names, it will return the quantity value, because the average of each total is that value. When the measure is evaluated for the Total row, however, it can then calculate the average of all the total values; see Figure 6-13.

If you want to calculate the maximum or minimum total quantity value, you can use MAXX and MINX, respectively, in the same way we have used AVERAGEX.

Q42. How Do I Get My Totals to Calculate Correctly?

On the Microsoft Power BI Ideas Forum (`https://ideas.powerbi.com/ideas/`), there has been a long-running request, as follows:

It would be great if Grand Totals on a Matrix or Table would add up correctly to the sum of the rows when you include MEASURES in the visual. Currently, the total on a measure consists of the measure function itself, instead of the sum of the rows above. PLEASE give us the option to toggle totals to be a simple aggregate of the rows above also.

As we identified in Q39, "Why Doesn't My Measure Work?" the calculation on the Total row of a table or matrix visual is the same DAX expression that was applied to each individual row of the visual but only this time is applied in a different filter context where filters have been removed. Therefore, the Total row is not the sum of the values in each row above. People often complain that this shouldn't be the case and that table visuals should always sum these values in the Total row accordingly. However, the fact that people don't understand the context of evaluations in the Total row of table and matrix visuals is not the fault of DAX. We're going to explore two examples of how the calculation on the Total row can appear to be incorrect, when in fact, it's simply a case of identifying the filter context in which the total row is being evaluated and taking steps to ensure that the calculation holds true to what we expect to see.

In our first example, we'll create these two measures that we can see evaluated in Figure 6-14:

```
Total Qty =
SUM ( Sales[QUANTITY] )

Total Qty Plus 100 =
SUM ( Sales[QUANTITY] ) + 100
```

WINE	Total Qty	Total Qty Plus 100
Bordeaux	9,183	9,283
Champagne	22,854	22,954
Chardonnay	23,737	23,837
Chenin Blanc	16,419	16,519
Chianti	22,130	22,230
Grenache	25,942	26,042
Lambrusco		100
Malbec	16,128	16,228
Merlot	18,141	18,241
Piesporter	19,741	19,841
Pinot Grigio	16,560	16,660
Rioja	21,546	21,646
Sauvignon Blanc	20,012	20,112
Shiraz	19,517	19,617
Total	**251,910**	**252,010**

Figure 6-14. *The Total row calculation is not correct for "Total Qty Plus 100"*

We can see that "Total Qty Plus 100" does not return the correct value on the Total row. Clearly, what has happened here is that 100 has been added to the grand total of the sum of quantity, exactly according to the calculation of the measure. The way we can fix this is to use the SUMX function which is usually integral to ensuring the Total row shows the sum of the values in each individual row of the table or matrix visual. What we must do is generate a measure that will require just two elements:

1. Referencing a table (real or virtual) that holds the categorical columns in the matrix or table visual where the total row is being evaluated. Therefore, in our example, we must reference the Products dimension because it holds the WINE column that categorizes the data in the table visual.

2. The SUMX function passed over this table to aggregate the original measure. Therefore, we must use SUMX to iterate the Products dimension to sum the "Total Qty Plus 100" measure.

This is the measure we must generate:

```
Total Qty Plus 100 Correct =
SUMX ( Products, [Total Qty Plus 100] )
```

In the "Total Qty Plus 100 Correct" measure, the "Total Qty Plus 100" nested measure, when passed into the Products dimension by SUMX, will calculate correctly for each row in the Products table (we explored this concept in Q41, "How Can I Calculate the Average of the Totals?"). When the measure evaluates the Total row, filters are removed from the Products dimension, and therefore, the measure will sum the individual results of "Total Qty Plus 100" for each row, resulting in the correct total. You can see in Figure 6-15 that the Total row now sums the values mentioned.

WINE	Total Qty	Total Qty Plus 100	Total Qty Plus 100 Correct
Bordeaux	9,183	9,283	9,283
Champagne	22,854	22,954	22,954
Chardonnay	23,737	23,837	23,837
Chenin Blanc	16,419	16,519	16,519
Chianti	22,130	22,230	22,230
Grenache	25,942	26,042	26,042
Lambrusco		100	100
Malbec	16,128	16,228	16,228
Merlot	18,141	18,241	18,241
Piesporter	19,741	19,841	19,841
Pinot Grigio	16,560	16,660	16,660
Rioja	21,546	21,646	21,646
Sauvignon Blanc	20,012	20,112	20,112
Shiraz	19,517	19,617	19,617
Total	**251,910**	**252,010**	**253,310**

Figure 6-15. *Using the "Total Qty Plus 100 Correct" measure*

If we want to be pedantic, because we require only the names of the wines sitting in the WINE column of the Products dimension for SUMX to iterate, we could re-write the "Total Qty Plus 100 Correct" measure using the VALUES function to retrieve just the wine name values, rather than referencing the entire Products dimension:

```
Total Qty Plus 100 Correct =
SUMX ( VALUES ( Products[WINE] ), [Total Qty Plus 100] )
```

The SUMMARIZE function would do the same job:

```
Total Qty Plus 100 Correct =
SUMX ( SUMMARIZE ( Products, Products[WINE] ), [Total Qty Plus 100] )
```

Using SUMMARIZE in this way allows us also to build correct totals when the matrix or table visual holds columns from more than one table, for instance, from the Products dimension and the SalesPeople dimension. For example, if we now build a matrix visual that includes the WINE column from the Products dimension and the SALESPERSON column from the SalesPeople dimension, our "Total Qty Plus 100 Correct" measure will now not work, Figure 6-16.

WINE	Total Qty	Total Qty Plus 100	Total Qty Plus 100 Correct
⊟ **Bordeaux**	**9,183**	**9,283**	**9,283**
Abel	1,472	1,572	1,572
Blanchet	894	994	994
Charron	2,205	2,305	2,305
Denis	1,057	1,157	1,157
Leblanc	2,745	2,845	2,845
Reyer	810	910	910
⊟ **Champagne**	**22,854**	**22,954**	**22,954**
Abel	4,320	4,420	4,420
Blanchet	4,325	4,425	4,425
Charron	3,611	3,711	3,711
Denis	3,696	3,796	3,796
Leblanc	3,369	3,469	3,469
Reyer	3,533	3,633	3,633
⊟ **Chardonnay**	**23,737**	**23,837**	**23,837**
Abel	2,690	2,790	2,790
Blanchet	1,629	1,729	1,729
Charron	4,522	4,622	4,622
Denis	5,878	5,978	5,978
Leblanc	3,493	3,593	3,593
Reyer	5,525	5,625	5,625
Total	**55,774**	**55,874**	**56,074**

Figure 6-16. *Adding in the Salespeople column to the matrix and the "Total Qty Plus 100 Correct" measure is now not correct*

We need to ensure that the table that SUMX iterates includes both the WINE column and the SALESPERSON column, but these columns sit in different dimensions. If we use SUMMARIZE, we can build a virtual table that mimics the columns in the rows of the matrix and then pass SUMX over this virtual table:

```
Total Qty Plus 100 Correct #2 =
SUMX (
    SUMMARIZE ( Sales, Products[WINE],
            SalesPeople[SALESPERSON] ),
    [Total Qty Plus 100] )
```

You can now understand that by using SUMMARIZE in this way, we can build virtual tables that combine columns from any dimensions, according to the structure of the matrix or table visual.

We now have the correct subtotals and grand total; see Figure 6-17.

WINE	Total Qty	Total Qty Plus 100	Total Qty Plus 100 Correct #2
⊟ **Bordeaux**	**9,183**	**9,283**	**9,783**
Abel	1,472	1,572	1,572
Blanchet	894	994	994
Charron	2,205	2,305	2,305
Denis	1,057	1,157	1,157
Leblanc	2,745	2,845	2,845
Reyer	810	910	910
⊟ **Champagne**	**22,854**	**22,954**	**23,454**
Abel	4,320	4,420	4,420
Blanchet	4,325	4,425	4,425
Charron	3,611	3,711	3,711
Denis	3,696	3,796	3,796
Leblanc	3,369	3,469	3,469
Reyer	3,533	3,633	3,633
⊟ **Chardonnay**	**23,737**	**23,837**	**24,337**
Abel	2,690	2,790	2,790
Blanchet	1,629	1,729	1,729
Charron	4,522	4,622	4,622
Denis	5,878	5,978	5,978
Leblanc	3,493	3,593	3,593
Reyer	5,525	5,625	5,625
Total	**55,774**	**55,874**	**57,574**

Figure 6-17. *The "Total Qty Plus 100 Correct" is now correct*

In the second example, we meet a common mistake into which newbies to DAX often fall. They understand that DAX measures must include an aggregation and so wrap everything inside SUM. For instance, they may want to calculate the total sales value for each product by multiplying the price in the Products table with the quantity in the Sales table:

```
Total Revenue Wrong =
SUM(Products[PRICE PER CASE]) * SUM(Sales[QUANTITY])
```

In Figure 6-18, we can see the evaluation of this measure along with the components expressed as separate measures, "Sum Price" and "Total Qty":

```
Sum Price =
SUM(Products[PRICE PER CASE])
```

```
Total Qty =
SUM(Sales[QUANTITY])
```

Note that the "Total Revenue" measure is correct for each product but is incorrect in the Total row.

WINE	Sum Price	Total Qty	Total Revenue Wrong
Bordeaux	$75	9,183	$688,725
Champagne	$100	22,854	$2,285,400
Chardonnay	$75	23,737	$1,780,275
Chenin Blanc	$50	16,419	$820,950
Chianti	$40	22,130	$885,200
Grenache	$30	25,942	$778,260
Lambrusco	$20		
Malbec	$85	16,128	$1,370,880
Merlot	$39	18,141	$707,499
Piesporter	$30	19,741	$592,230
Pinot Grigio	$30	16,560	$496,800
Rioja	$45	21,546	$969,570
Sauvignon Blanc	$40	20,012	$800,480
Shiraz	$78	19,517	$1,522,326
Total	**$737**	**251,910**	**$185,657,670**

Figure 6-18. *The "Total Revenue" measure is incorrect in the Total row*

Here, in the evaluation of each product, "Sum Price" is the price of each product because each product has been filtered in turn. When the evaluation arrives at the total row, however, all filters have been removed from the Products table, and therefore "Sum Price" sums the price values for all the products. Clearly what has happened here is that the Total row calculation is exactly the same as the calculation for each product, that is, the "Sum Price" value is multiplied by the "Sum Qty" value. It's just that, on the Total row, this is **$737 * 251,910**.

We can apply the same technique for calculating the total row correctly as we did before. That is, we must first reference a table (real or virtual) that holds the categorical columns in the matrix or table visual where the total row is being evaluated and secondly use SUMX passed over this table to aggregate the original measure. This is the measure we must generate:

```
Total Revenue Correct =
SUMX( Products, [Total Revenue Wrong] )
```

In Figure 6-19, we can see that we now get the correct values for each product and the correct value in the Total row.

WINE	Sum Price	Total Qty	Total Revenue Wrong	Total Revenue Correct
Bordeaux	$75	9,183	$688,725	$688,725
Champagne	$100	22,854	$2,285,400	$2,285,400
Chardonnay	$75	23,737	$1,780,275	$1,780,275
Chenin Blanc	$50	16,419	$820,950	$820,950
Chianti	$40	22,130	$885,200	$885,200
Grenache	$30	25,942	$778,260	$778,260
Lambrusco	$20			
Malbec	$85	16,128	$1,370,880	$1,370,880
Merlot	$39	18,141	$707,499	$707,499
Piesporter	$30	19,741	$592,230	$592,230
Pinot Grigio	$30	16,560	$496,800	$496,800
Rioja	$45	21,546	$969,570	$969,570
Sauvignon Blanc	$40	20,012	$800,480	$800,480
Shiraz	$78	19,517	$1,522,326	$1,522,326
Total	**$737**	**251,910**	**$185,657,670**	**$13,698,595**

Figure 6-19. *The "Total Revenue Correct" measure is now correct in the Total row*

However, if you need to calculate the total revenue by multiplying the QUANTITY column in the Sales table with the PRICE PER CASE column in the Products dimension, there is a better approach that doesn't require nesting measures inside SUMX. For this, you must read the answer to the next question.

Q43. How Can I Include Values from Dimensions in Calculations in the Fact Table?

One reason why people are reluctant to adopt star schemas when generating data models is that they are at a loss as to how to "reach" values stored in dimensions. The reason they must do this is to perform calculations that use data from both dimensions and from the fact table.

For example, in our fact table, we may not have a "Revenue" column. This shouldn't be a problem because we can calculate the revenue value. We have the QUANTIY value in the Sales fact table, and in the Products dimension, we have recorded the PRICE PER CASE values for each product. Clearly, using a calculated column, all we must do is multiply these two values. But how do we reference the PRICE PER CASE column in the calculation? As you can see in Figure 6-20, we will get an error if we attempt this calculated column:

```
REVENUE = Sales[QUANTITY] * Products[PRICE PER CASE]
```

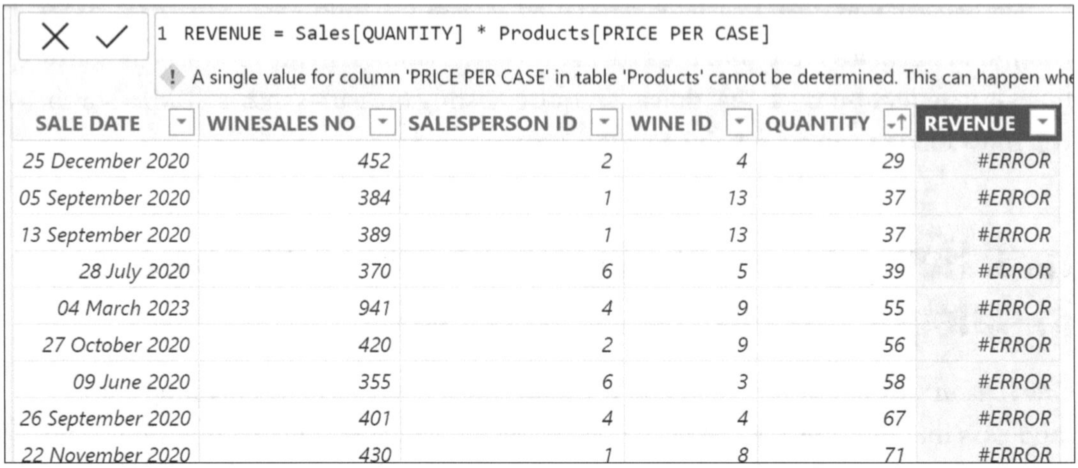

Figure 6-20. *Multiplying QUANTITY by PRICEPERCASE doesn't work*

The problem is that the expression does not know from which row in the Products dimension the price must be found. To solve this problem, we can use the RELATED function to retrieve the correct price. As its name suggests, this function can only be used if there is a many-to-one relationship between tables. The RELATED function will retrieve values from the one side of the relationship to be populated into many. Therefore, we can use it to combine values from dimensions with values from the fact table. RELATED can be used in much the same way that you would use the VLOOKUP or XLOOKUP functions in Excel.

We can rewrite the expression in the calculated column as follows:

```
REVENUE = Sales[QUANTITY] * RELATED ( Products[PRICE PER CASE] )
```

You can see in Figure 6-21 that now we have been able to include the value from the Products dimension with a value from the Sales fact table.

✕ ✓	1 REVENUE = Sales[QUANTITY] * RELATED (Products[PRICE PER CASE])				

SALE DATE ▾	WINESALES NO ▾	SALESPERSON ID ▾	WINE ID ▾	QUANTITY ▾↑	REVENUE ▾
25 December 2020	452	2	4	29	£2,465
15 September 2020	384	1	13	37	£2,886
3 September 2020	389	1	13	37	£2,886
28 July 2020	370	6	5	39	£1,170
04 March 2023	941	4	9	55	£2,145
27 October 2020	420	2	9	56	£2,184
09 June 2020	355	6	3	58	£4,350
16 September 2020	401	4	4	67	£5,695
22 November 2020	430	1	8	71	£2,130
01 November 2020	422	5	11	77	£3,465
08 October 2019	232	6	13	78	£6,084
12 January 2020	262	6	11	79	£3,510

Figure 6-21. *Using RELATED to retrieve values from dimensions*

The REVENUE calculated column may seem like a solution for calculating total sales values in a report on the canvas, especially for those new to DAX. It can be used as an implicit measure in visuals to determine the total revenue for customers or products, for example.

However, this may not be the best approach. It's possible that the REVENUE column doesn't fit well within the fact table as a calculated column. Consider this: firstly, the calculated column needs to be evaluated for *every row* in the Sales fact table and, secondly, *recalculated* whenever the data is refreshed. This can be a significant processing load, especially if the fact table contains millions of rows.

Thus, the question arises: if a calculated column is not suitable for the revenue calculation, what alternative should be used instead? We should always use measures if the calculation is not bound to the rows in a table. We're not necessarily interested in individual transaction values that sit in each row; we could have millions of these. More likely we want to analyze the total revenue value when it's aggregated against values from dimensions. Therefore, the total revenue calculation sits more comfortably in our data model as a measure.

The calculation for the measure is a little more challenging than the calculation in the column. It requires the use of the RELATED function along with SUMX:

```
Total Revenue =
SUMX(Sales, Sales[QUANTITY] *
            RELATED ( Products[PRICE PER CASE] ))
```

We can now use this measure to calculate the total revenue as shown in Figure 6-22.

WINE	Total Revenue
Bordeaux	$688,725
Champagne	$2,285,400
Chardonnay	$1,780,275
Chenin Blanc	$820,950
Chianti	$885,200
Grenache	$778,260
Malbec	$1,370,880
Merlot	$707,499
Piesporter	$592,230
Pinot Grigio	$496,800
Rioja	$969,570
Sauvignon Blanc	$800,480
Shiraz	$1,522,326
Total	**$13,698,595**

Figure 6-22. *Generating a Total Revenue measure using RELATED*

The "Total Revenue" measure, evaluated in Figure 6-22, uses SUMX to iterate each row in the Sales table that is filtered to hold the sales for each product, for example, for "Bordeaux" in the first evaluation. In the Sales table, the RELATED function finds the PRICE PER CASE value in the Products table for the product in the current row of the Sales table. It then multiplies this by the QUANTITY value in the current row of the Sales table. SUMX then sums the results of these row-level calculations. In Figure 6-23, we show the expression that SUMX sums *in memory* for each row in the Sales table.

The calculation performed in memory

```
SUMX ( Sales, Sales[QUANTITY] * RELATED( Products[PRICE PER CASE] ))
```

SALE DATE	WINESALES NO	SALESPERSON ID	WINE ID	QUANTITY	
11 January 2023	906	2	1	296	£22,200
19 October 2022	863	2	1	142	£10,650
12 May 2022	762	2	1	314	£23,550
08 May 2022	758	2	1	142	£10,650
29 April 2023	978	5	1	265	£19,875
13 April 2023	968	5	1	101	£7,575
12 February 2023	929	5	1	297	£22,275
13 November 2022	873	5	1	342	£25,650
07 October 2022	856	5	1	342	£25,650
04 September 2022	830	5	1	172	£12,900
05 August 2022	812	5	1	172	£12,900
07 August 2021	643	5	1	172	£12,900
31 July 2021	639	5	1	342	£25,650
17 September 2022	844	1	1	466	£34,950
06 March 2022	737	1	1	130	£9,750
01 July 2021	626	1	1	410	£30,750

Figure 6-23. *The evaluation of "Total Revenue" using SUMX and RELATED*

Therefore, all the while you have a star schema, you can use the RELATED function to move values from dimensions into the fact table and perform calculations across these tables.

Q44. How Do I Bin Measures into Numeric Ranges?

So, for example, you want to uncover how many customers have a "Total Qty" value that falls between, for example, 1,000 and 1,499. Take a look at the visual depicted on the left side of Figure 6-24. It tells us there are two customers, "Chalottesville & Co" and "Kassel & Sons," whose "Total Qty" values fall between these parameters.

CUSTOMER NAME	Total Qty ▲	^	Range ▲	No. of Customers with This Total Qty
Cheney & Co	782		500 to 999	1
Charlottesville & Co	1,258		1,000 to 1,499	2
Kassel & Sons	1,272		1,500 to 1,999	13
Port Hammond Bros	1,540		2,000 to 2,499	14
Rhodes Ltd	1,604		2,500 to 2,999	11
Palo Alto Ltd	1,662		3,000 to 3,499	16
St. Leonards Ltd	1,690		3,500 and more	27
Eilenburg Ltd	1,759		**Total**	**84**
Morristown & Co	1,774			
Victoria Ltd	1,775			
Yokohama & Co	1,810			
Hervey Bay Bros	1,927			
Barstow Ltd	1,934			
Waterbury Ltd	1,939			
Chandler & Sons	1,980			
Westminster Bros	1,986	∨		
Total	**251,910**			

Figure 6-24. *Binning "Total Qty" into numeric ranges*

If this is the kind of analysis you'd like to achieve, the first step is to create a parameter table that defines your desired ranges. You can accomplish this by using the "**Enter Data**" button found on the **Home** tab. In Figure 6-25, you will see the "Bins for Measures" parameter table that we've created. Similar to all parameter tables, it is not linked to any other tables within the model.

Range ▼	Min Value ▼	Max Value ▼	Sort ▼
500 to 999	500	999	1
1,000 to 1,499	1000	1499	2
1,500 to 1,999	1500	1999	3
2,000 to 2,499	2000	2499	4
2,500 to 2,999	2500	2999	5
3,000 to 3,499	3000	3499	6
3,500 and more	3500	9999999	7

Figure 6-25. *The "Bins for Measures" parameter table*

Now you have your parameter table, you can filter the Customers dimension where each customer's total quantity value falls between each parameter and count the rows of the filtered table. This is the measure that will do this:

```
No. of Customers with This Total Qty =
COUNTROWS (
        FILTER (
          Customers,
          [Total Qty] >= MIN('Bins for Measures'[Min Value])
              && [Total Qty] <=
                  MAX('Bins for Measures'[Max Value]) ))
```

You can then place a table visual on your canvas and populate it with the "Range" column from the "Bins for Measures" table. Next, place the "No. of Customers with this Total Qty" measure into this table as in Figure 6-24.

Q45. Why Does Filtering Zeros Also Include Blank Values?

Let's first peruse an example of this behavior. Consider the data shown in Figure 6-26.

CUSTOMER NAME	Total Revenue
Acme & Sons	
Bloxon Bros.	
Jones Ltd	
Sainsbury's	
Smith & Co	$0
Cheney & Co	$41,660
Port Hammond Bros	$50,170
Kassel & Sons	$67,439
Charlottesville & Co	$77,190
Rhodes Ltd	$88,086
Victoria Ltd	$93,788
Palo Alto Ltd	$95,801

Figure 6-26. *Blank and zero values for the "Total Revenue" measure in a table visual*

As well as values for the "Total Revenue" measure, there are also both blank values and a zero value. Let's now calculate the number of zero values returned by the "Total Revenue" measure, which should return **1**:

```
No. of Customers with Zero Sales =
COUNTROWS ( FILTER ( Customers, [Total Revenue] = 0 ) )
```

However, when placed inside a card visual, this measure returns **5**; see Figure 6-27.

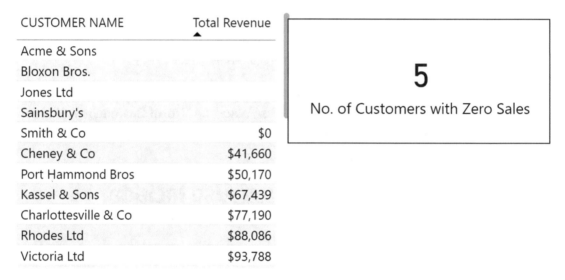

Figure 6-27. *The "No. of Customers with Zero Sales" measure returns 5*

The reason we get 5 here is because DAX converts blank values to 0 on the evaluation of a measure. Therefore, what we can conclude from this example is that DAX treats zeros and blank values as the same, and therefore, filtering zero values will also include blank values. So how can we find the number of zero sales? For this, we can use the ISBLANK function that returns TRUE if a blank is found. Therefore, this measure will return **4**.

```
No. of Customers with Blank Sales =
COUNTROWS ( FILTER ( Customers, ISBLANK( [Total Revenue] ) ))
```

We can now wrap ISBLANK inside the NOT function to eliminate blank values from the filter:

```
No. of Customers with Zero Sales=
COUNTROWS (
    FILTER ( Customers, NOT (
      ISBLANK ( [Total Revenue]  ) )
          && [Total Revenue]  = 0 ) )
```

This measure returns the correct value; see Figure 6-28.

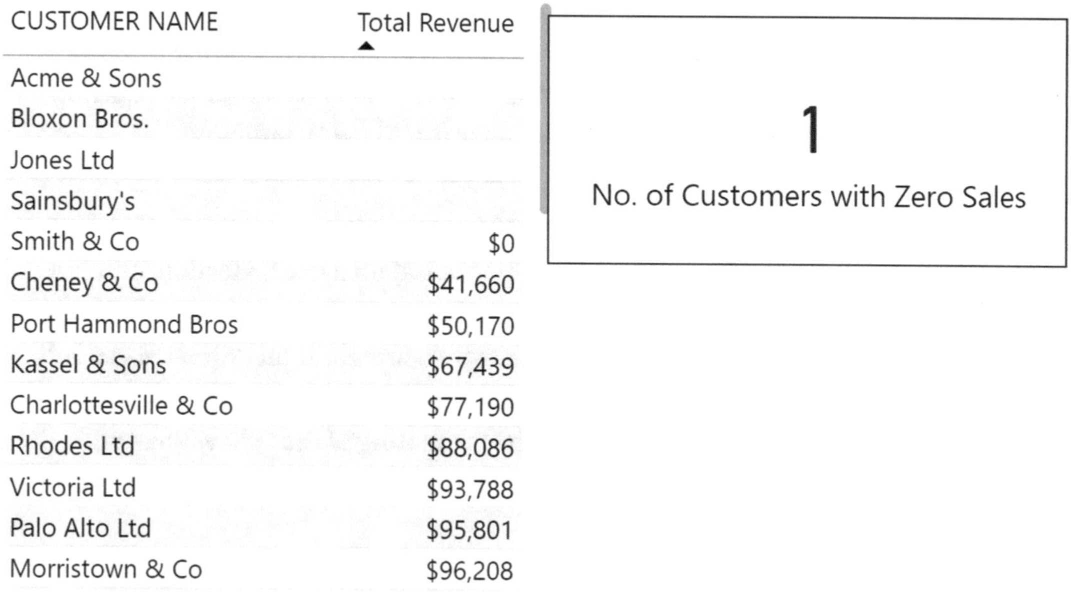

CUSTOMER NAME	Total Revenue ▲
Acme & Sons	
Bloxon Bros.	
Jones Ltd	
Sainsbury's	
Smith & Co	$0
Cheney & Co	$41,660
Port Hammond Bros	$50,170
Kassel & Sons	$67,439
Charlottesville & Co	$77,190
Rhodes Ltd	$88,086
Victoria Ltd	$93,788
Palo Alto Ltd	$95,801
Morristown & Co	$96,208

> **1**
>
> No. of Customers with Zero Sales

Figure 6-28. *The "No. of Customers with Zero Sales" measure now returns 1*

There is an alternative approach to calculating the number of customers with zero sales. You can use the "strictly equal to" operator which is the double equals sign (==). This will return TRUE if the two arguments on either side of the operator are the same:

```
No. of Customers with Zero Sales v2 =
COUNTROWS(FILTER(Customers,[Total Revenue]==0))
```

You'll appreciate that this measure is greatly more succinct than using NOT and ISBLANK.

Either way, next time you filter zero values, remember that unless you use ISBLANK to weed out blank values, DAX will treat blank values and zero values the same.

Q46. Can I Have Different Numeric Formatting on Values Returned by the Same Measure?

Yes, you can! In the April 2023 release of Power BI Desktop, the "Dynamic" format option became available for measures. This option allows you to apply different numeric formats to values returned within the same measure.

Note   At the time of writing, the "Dynamic format strings" are a preview feature. To turn on Preview features, use the **File** tab, **Options and settings** and **Options**. Then find the **Preview features** category under the GLOBAL settings.

To explore an example of where this would be a necessity, we can look at designing a measure that in itself returns different measures based on a user selection, and the measures selected require different formatting. In Figure 6-29, we are showing three instances of the same table visual, showing the wine names and the "Show Measure" measure. This measure will in itself display a different measure in the table visual depending on the slicer selection. Note how the formatting of the "Show Measure" measure changes accordingly. This was achieved using the "Dynamic" format option on the Measure tools tab.

WINE	Show Measure	Select Measure		WINE	Show Measure	Select Measure
Bordeaux	3.65%	■ % Grand Total		Bordeaux	9,183	☐ % Grand Total
Champagne	9.07%	☐ Total Qty		Champagne	22,854	■ Total Qty
Chardonnay	9.42%	☐ Total Revenue		Chardonnay	23,737	☐ Total Revenue
Chenin Blanc	6.52%			Chenin Blanc	16,419	
Chianti	8.78%			Chianti	22,130	
Grenache	10.30%			Grenache	25,942	
Malbec	6.40%			Malbec	16,128	
Merlot	7.20%			Merlot	18,141	
Piesporter	7.84%			Piesporter	19,741	
Pinot Grigio	6.57%			Pinot Grigio	16,560	
Rioja	8.55%			Rioja	21,546	
Sauvignon Blanc	7.94%			Sauvignon Blanc	20,012	
Shiraz	7.75%			Shiraz	19,517	
Total	**100.00%**			**Total**	**251,910**	

WINE	Show Measure	Select Measure
Bordeaux	$688,725	☐ % Grand Total
Champagne	$2,285,400	☐ Total Qty
Chardonnay	$1,780,275	■ Total Revenue
Chenin Blanc	$820,950	
Chianti	$885,200	
Grenache	$778,260	
Malbec	$1,370,880	
Merlot	$707,499	
Piesporter	$592,230	
Pinot Grigio	$496,800	
Rioja	$969,570	
Sauvignon Blanc	$800,480	
Shiraz	$1,522,326	
Total	**$13,698,595**	

Figure 6-29. *The "Show Measure" measure changes its formatting depending on the measure selected by the slicer*

If you would like to follow along with this example, first you will need to generate a parameter table that holds the measure names and a value assigned to each measure (refer to Q44 for building parameter tables). Our parameter table is named "Measure Select"; see Figure 6-30.

Select Measure ▾	Value ▾
% Grand Total	1
Total Qty	2
Total Revenue	3

Figure 6-30. *The "Measure Select" parameter table*

In the "Select Measure" column of the "Measure Select" table, you don't need to match the text values to the names of the measures exactly – it's the "Value" that will be used to drive which measure displays in the visual. On the report canvas, you must put a slicer holding the "Select Measure" column from the "Measure Select" table. Now create this measure (substituting your measure names in the SWITCH function):

```
Show Measure =
SWITCH (
    SELECTEDVALUE ( 'Measure Select'[Value] ),
    1, [%GT Total Qty],
    3, [Total Revenue],
    2, [Total Qty])
```

Now for the dynamic formatting, with the "Show Measure" measure selected, on the **Measure tools** tab, select **Dynamic** from the **Format** dropdown; see Figure 6-31.

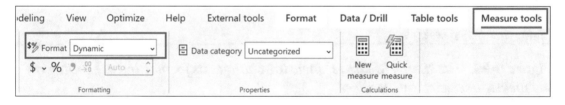

Figure 6-31. *Select "Dynamic" from the Format dropdown*

In the Format formula bar, type a DAX expression that will apply a format string to each of the different values being returned by your measure. You can see in Figure 6-32 that the expression is similar to the "Show Measure" expression; the difference is that we are applying different format strings depending on the slicer selection. The format strings use zero (0) as placeholders for digits in the values being returned.

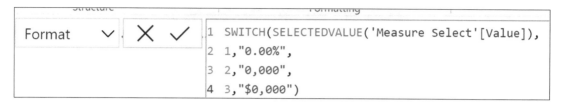

Format ∨ ✕ ✓	1 SWITCH(SELECTEDVALUE('Measure Select'[Value]),
	2 1,"0.00%",
	3 2,"0,000",
	4 3,"$0,000")

Figure 6-32. *The Dynamic formatting expression*

If you need to edit the dynamic formatting, select the measure, and then select "Format" in the dropdown on the left of the formula bar, as shown in Figure 6-32.

You may, in the past, have used the FORMAT function to achieve a similar outcome. However, the great advantage of using dynamic format strings is that they allow you to retain numeric data types on the values returned by measures instead of converting these to strings. This is particularly useful for charts that rely on numeric values. By utilizing dynamic format strings, measures can maintain their data type and be assigned different format strings based on the context.

Q47. How Can I Calculate the Mode of Values?

The reason this question arises is that, unlike Excel, DAX has no MODE function. Yes, it has AVERAGE, AVERAGEX, MEDIAN, and MEDIANX, but for some reason, we do not have MODE (or MODEX) in DAX. Therefore, we must calculate this ourselves.

For example, we might want to compare each salesperson's mode quantity sold to their average quantity sold. Perhaps, we must do this to identify the existence of outliers and establish those salespeople who may not sell consistent quantities in each sales transaction. You can see an example of these insights in Figure 6-33, where, because the data has been randomly generated, the mode and average are much the same.

SALESPERSON	Mode Qty	Avg Qty
Abel	285	256
Blanchet	295	249
Charron	142	253
Denis	246	260
Leblanc	273	268
Reyer	244	256
Total	**246**	**257**

Figure 6-33. *Calculating the mode quantity*

To calculate the mode in DAX, this is the measure you can use:

```
Mode Qty =
MAXX (
    TOPN (
        1,
        SUMMARIZE ( Sales, Sales[QUANTITY], "Occurrence",
                        COUNTROWS( Sales )),
        [Occurrence] ),
    Sales[QUANTITY] )
```

This measure generates a virtual table using SUMMARIZE that groups transactions by their quantity. It then counts the number of rows in each group, storing this value in the virtual column, "Occurrence." The TOPN function then finds the top value in the "Occurrence" column, and we can then use MAXX to return this top value.

If you want to calculate the mode of an expression, simulating the existence of a "MODEX" function, we can use similar code to the "Mode" measure. For example, we may want to calculate the mode of the Total Revenue measure. To do this, we must create a virtual table containing total revenue values for each sales transaction. We can then use SUMMARIZE to group by total revenue. This would be the code that would do this job:

```
Mode for Total Revenue =
MAXX (
    TOPN (
        1,
```

```
    SUMMARIZE (
  ADDCOLUMNS ( VALUES ( Sales[WINESALES NO] ), "Revenue", [Total
  Revenue]),
          [Revenue],
          "Occurrence", COUNTROWS ( Sales )
      ),
      [Occurrence]
  ),
  [Revenue]   )
```

To understand the "Mode for Total Revenue" measure, we can dissect each table expression within it. For example, this code

```
ADDCOLUMNS ( VALUES ( Sales[WINESALES NO] ), "Revenue",
                                    [Total Revenue]),
```

generates a virtual table comprising the WINESALES NO column from the Sales table and the "Total Revenue" values, in the "Revenue" column of the virtual table; see Figure 6-34.

WINESALES NO	Revenue
1	£10,800
2	£6,600
3	£9,000
4	£16,200
5	£22,698
6	£4,770
7	£7,680
8	£10,600
9	£9,040
10	£8,853
11	£10,215
12	£4,140

Figure 6-34. *Virtual table generated by ADDCOLUMNS*

Now this slice of DAX

```
SUMMARIZE (
        ADDCOLUMNS ( VALUES ( Sales[WINESALES NO] ), "Revenue",
                                        [Total Revenue] ),
            [Revenue],
            "Occurrence", COUNTROWS ( Sales )  )
```

uses the table generated by ADDCOLUMNS to group by revenue so that we can calculate how many rows there are in the Sales table that contain each revenue value; see Figure 6-35.

Revenue ▾	Occurrence ↴
£12,900	7
£4,080	6
£7,950	6
£13,800	5
£8,970	5
£5,460	5
£12,300	5
£8,850	5
£10,650	5
£9,000	5
£10,800	5

Figure 6-35. *The virtual table generated by SUMMARIZE*

Finally, this code

```
TOPN (
        1,
        SUMMARIZE (
          ADDCOLUMNS ( VALUES ( Sales[WINESALES NO] ), "Revenue",
                                          [Total Revenue] ),
              [Revenue],
              "Occurrence", COUNTROWS ( Sales ) ),
          [Occurrence] )
```

finds the top value in the "Occurrence" column as depicted in Figure 6-36.

Revenue	Occurrence
£12,900	7

Figure 6-36. *The table generated by TOPN*

Finally, the MAXX function returns the "Revenue" value shown in Figure 6-36. As there is only one row returned by TOPN for MAXX to iterate, you could use MINX or SUMX here. However, if the result is bimodal, SUMX will sum the two most frequent occurrences. Using MAXX means that at least it will return the largest of the mode values. The results of the "Mode for Total Revenue" measure are shown in Figure 6-37, along with the "Avg Revenue."

SALESPERSON	Mode for Total Revenue	Avg Revenue
Abel	$12,120	$13,525
Blanchet	$8,850	$12,770
Charron	$8,970	$13,967
Denis	$34,728	$14,636
Leblanc	$15,900	$15,340
Reyer	$5,460	$13,544
Total	**$12,900**	**$13,992**

Figure 6-37. *Calculating the mode Total Revenue*

Therefore, although there is no MODE or MODEX function in DAX, we can calculate these insights for ourselves.

CHAPTER 7

Customers and Products

We've labeled this chapter as shown because the questions within it typically pertain to gaining insights into your customer base or the products or services you offer. Nevertheless, "customers and products" serves as a broad term encompassing any entities within your data model. The answers we provide here are equally relevant to, for instance, sales regions or salespeople and can be adapted accordingly.

Q48. How Can I Rank or Find TopN Customers or Products?

If you want to rank your customers or products by, for example, total revenue, you can use the RANKX function as follows:

```
Rank Customers =
RANKX ( ALLSELECTED ( Customers ), [Total Revenue] )
```

In this expression, the RANKX function iterates the rows in the Customers table for the customers shown in the visual, calculating the total revenue value for each customer and ranking customers accordingly; see Figure 7-1.

© Alison Box 2023
A. Box, *A Power BI Compendium*, https://doi.org/10.1007/978-1-4842-9765-0_7

CUSTOMER NAME	Total Revenue ▼	Rank Customers
Issaquah & Co	$338,880	1
Concord Ltd	$317,212	2
Castle Rock Ltd	$291,620	3
Ballard & Sons	$264,262	4
Fort Worth Ltd	$262,490	5
Busan & Co	$262,410	6
Germantown & Co	$260,441	7
Burningsuit Ltd	$257,055	8
Branch Ltd	$253,026	9
Liverpool & Sons	$249,550	10
Clifton Ltd	$246,915	11
Landstuhl Ltd	$243,246	12
Lavender Bay Ltd	$241,784	13

Figure 7-1. *The RANKX function ranks customers descending by default*

The default is to rank entities in descending order by value, so the highest total revenue is ranked **1**. If you want to sort ascending by value, you can edit the measure as shown:

```
Rank Customers =
RANKX ( ALLSELECTED ( Customers ), [Total Revenue],,ASC )
```

However, as you can see in Figure 7-2, customers with no sales are ranked "**1**," and, therefore, the first customer with a revenue value is now ranked "**6**."

CUSTOMER NAME	Total Revenue	Rank Customers
Acme & Sons		1
Bloxon Bros.		1
Jones Ltd		1
Sainsbury's		1
Smith & Co		1
Cheney & Co	$41,660	6
Port Hammond Bros	$50,170	7
Kassel & Sons	$67,439	8
Charlottesville & Co	$77,190	9
Rhodes Ltd	$88,086	10
Victoria Ltd	$93,788	11
Palo Alto Ltd	$95,801	12
Morristown & Co	$96,208	13
Black Ltd	$98,080	14

Figure 7-2. *You can change the RANKX function to rank customers ascending*

To resolve this dilemma, you can edit the measure as follows:

```
Rank Customers ASC =
IF ( [Total Revenue],
    RANKX ( FILTER ( ALL ( Customers ), [Total Revenue] ),
            [Total Revenue],, ASC ) )
```

In the edited version of the ranking measure, we first test for the presence of a total revenue value. This will eliminate the blank total revenue values in the visual. The FILTER function then filters those customers who have total revenue values, and, therefore, RANKX ranks customers accordingly; the customer with the lowest revenue is now ranked "**1**."

In Figure 7-3, we are now ranking our customers ascending by rank value and have eliminated customers with no sales from the ranking.

CUSTOMER NAME	Total Revenue ▲	Rank Customers ASC
Cheney & Co	$41,660	1
Port Hammond Bros	$50,170	2
Kassel & Sons	$67,439	3
Charlottesville & Co	$77,190	4
Rhodes Ltd	$88,086	5
Victoria Ltd	$93,788	6
Palo Alto Ltd	$95,801	7
Morristown & Co	$96,208	8
Black Ltd	$98,080	9
Beaverton & Co	$99,240	10
Total	**$13,698,595**	**85**

Figure 7-3. *Eliminating customers with no sales from the ranking*

However, you may have noticed that the Total row has the highest rank value of **85** because it's the largest value. This value is not required, so we can prevent an evaluation in the Total row by using this code:

```
Rank Customers ASC =
IF ( [Total Revenue],
IF ( HASONEVALUE(Customers[CUSTOMER NAME] ),
    RANKX ( FILTER ( ALL ( Customers ), [Total Revenue] ),
            [Total Revenue],, ASC ) ) )
```

Therefore, ranking entities using the RANKX function is relatively straightforward to accomplish in DAX.

If your requirement is simply to analyze topN customers, for example, top three customers by total revenue, you can use the "TopN" visual level filter to do this. Consider Figure 7-4 where we have used this filter in a matrix visual to show the top three customers by the "Total Revenue" measure.

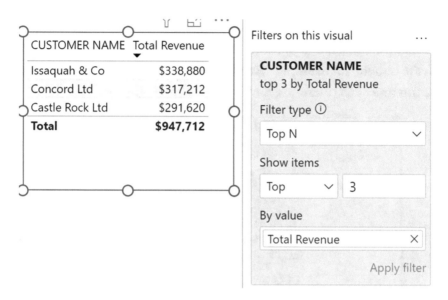

Figure 7-4. *Using the visual filter to find TopN*

However, it may be that you want to analyze the top three customers who purchased each product. Therefore, you might now include product names in the rows of the matrix; see Figure 7-5.

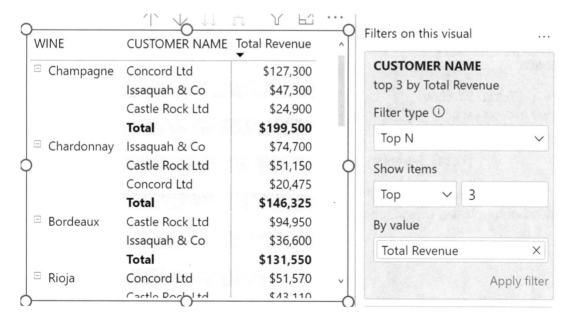

Figure 7-5. *Adding the product names to the matrix*

259

The trouble is that the matrix still shows the *same* top three customers across ALL products (if they purchased that product). What we want to show are the top three customers who bought, for example, "Bordeaux." If we remove the TopN filter, we can see who the top three customers are for "Bordeaux" in Figure 7-6.

WINE	CUSTOMER NAME	Total Revenue
Bordeaux	Castle Rock Ltd	$94,950
	Nizhny Novgorod & Co	$91,200
	Chatou & Co	$76,950
	Littleton & Sons	$60,750
	Shanghai Ltd	$52,500
	Burningsuit Ltd	$45,900
	Barstow Ltd	$38,700
	Issaquah & Co	$36,600
	Clifton Ltd	$33,375

Figure 7-6. The top three customers selling "Bordeaux"

So how do we get these values to show in the matrix visual alongside "Bordeaux"? For this, we can use the same DAX measure we used to rank the customers (this time ranking descending) and use a visual level filter to filter those customers ranked three or less.

```
IF ( [Total Revenue],
IF ( HASONEVALUE(Customers[CUSTOMER NAME] ),
    RANKX ( FILTER ( ALL ( Customers ), [Total Revenue] ),
            [Total Revenue] ) ) )
```

In Figure 7-7, you can see how using this measure in conjunction with a visual level filter is a simple way to find topN.

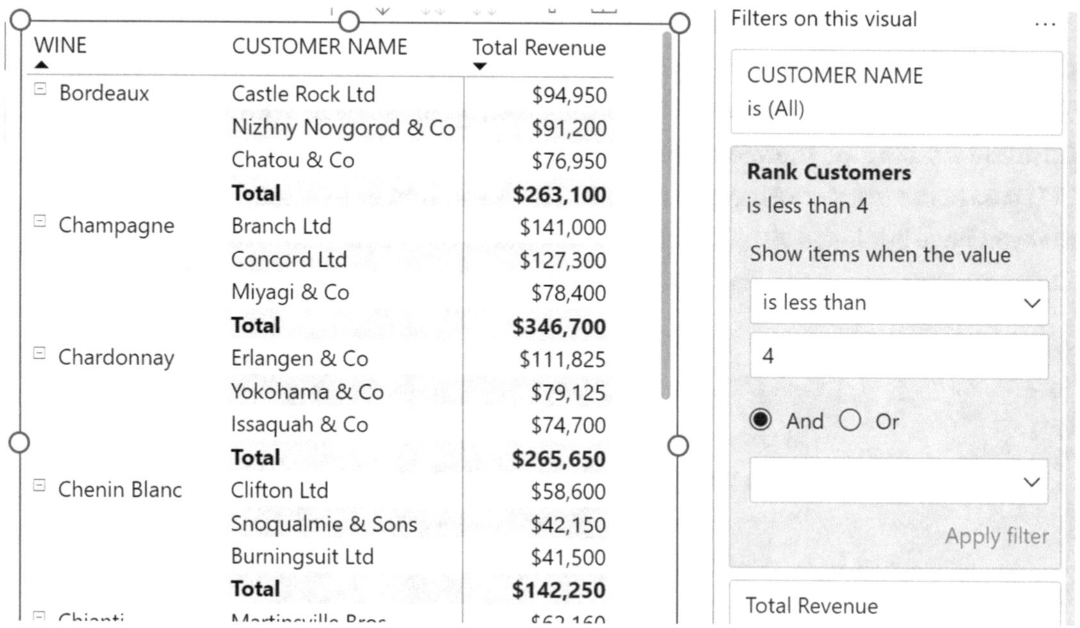

WINE ▲	CUSTOMER NAME	Total Revenue ▼
⊟ Bordeaux	Castle Rock Ltd	$94,950
	Nizhny Novgorod & Co	$91,200
	Chatou & Co	$76,950
	Total	**$263,100**
⊟ Champagne	Branch Ltd	$141,000
	Concord Ltd	$127,300
	Miyagi & Co	$78,400
	Total	**$346,700**
⊟ Chardonnay	Erlangen & Co	$111,825
	Yokohama & Co	$79,125
	Issaquah & Co	$74,700
	Total	**$265,650**
⊟ Chenin Blanc	Clifton Ltd	$58,600
	Snoqualmie & Sons	$42,150
	Burningsuit Ltd	$41,500
	Total	**$142,250**
⊟ Chianti	Martinsville Bros	$62,160

Filters on this visual   ...

CUSTOMER NAME
is (All)

Rank Customers
is less than 4

Show items when the value

is less than   ∨

4

◉ And ○ Or

∨

Apply filter

Total Revenue

Figure 7-7. *Finding top three customers by product*

So far, we've been exploring "topN" analysis. However, better insights into the best and worst performing customers or products can come when we analyze the topN percent value instead. This is the DAX to calculate the top 10 percent of customers by sales revenue:

```
Top 10 PC Customers =
VAR MyPercent =
    COUNTROWS ( FILTER ( ALL ( Customers ), [Total Revenue] ) ) * 0.1
VAR RankCusts =
    RANKX ( ALLSELECTED ( Customers ), [Total Revenue] )
VAR FindCusts =
    FILTER ( Customers, RankCusts <= MyPercent)
RETURN
    CALCULATE ( [Total Revenue], FindCusts )
```

Here, the "MyPerCent" variable finds the percent value to be used. For example, we have 84 customers with a total revenue, so this value will be 8.4. The "RankCusts" variable ranks the customers by Total Revenue. In "FindCusts" the FILTER function filters customers whose rank is less than or equal to 8.9. Finally, CALCULATE calculates the Total Revenue for the filtered customers.

When you put this code into a table visual, notice the Total row shows the grand total revenue and not the sum of the sales revenue for the top 10 percent of customers. This is because in the evaluation of the Total row, all filters have been removed from the Customers table and, therefore, the rank for the total revenue for all customers (which is "**1**") passes the test performed by FILTER. The resolution to this issue is simple; ensure that you have applied a visual level filter that filters out blank values for the "Top 10 PC Customers" measure; see Figure 7-8.

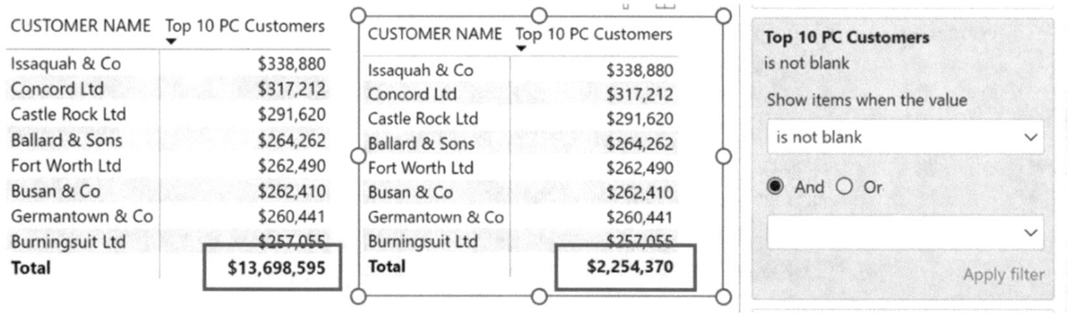

CUSTOMER NAME	Top 10 PC Customers
Issaquah & Co	$338,880
Concord Ltd	$317,212
Castle Rock Ltd	$291,620
Ballard & Sons	$264,262
Fort Worth Ltd	$262,490
Busan & Co	$262,410
Germantown & Co	$260,441
Burningsuit Ltd	$257,055
Total	**$13,698,595**

CUSTOMER NAME	Top 10 PC Customers
Issaquah & Co	$338,880
Concord Ltd	$317,212
Castle Rock Ltd	$291,620
Ballard & Sons	$264,262
Fort Worth Ltd	$262,490
Busan & Co	$262,410
Germantown & Co	$260,441
Burningsuit Ltd	$257,055
Total	**$2,254,370**

Top 10 PC Customers
is not blank

Show items when the value

is not blank

● And ○ Or

Apply filter

Figure 7-8. Calculating the top 10 percent of customers by sales revenue using a visual filter to calculate the Total row correctly

To calculate the bottom 10 percent, all that is required is to reverse the sort order of the RANKX function to ascending so that the lowest value is assigned "**1**" in the ranking order:

```
Bottom 10 PC Customers =
VAR MyPercent =
    COUNTROWS ( ALL ( Customers ) ) * 0.1
VAR RankCusts =
    RANKX (FILTER( ALLSELECTED ( Customers ),[Total Revenue]),
                                      [Total Revenue] ,,ASC )
VAR FindCusts =
    FILTER ( Customers, RankCusts <= MyPercent)
RETURN
  CALCULATE ( [Total Revenue], FindCusts )
```

You can see the results of this measure in Figure 7-9. However, you will notice again that we have issues with the Total row; it has a blank result. Just as before, this is because the Total row value is assigned a ranking order, just as the individual sales revenue values. Because this value will always have a greater ranking value than the individual sales values, the Total value fails the test supplied by the "FindCusts" variable.

CUSTOMER NAME	Bottom 10 PC Customers ▲
Cheney & Co	$41,660
Port Hammond Bros	$50,170
Kassel & Sons	$67,439
Charlottesville & Co	$77,190
Rhodes Ltd	$88,086
Victoria Ltd	$93,788
Palo Alto Ltd	$95,801
Morristown & Co	$96,208
Total	

Figure 7-9. *Calculating the bottom 10 percent of customers by sales revenue*

The solution to this problem is different from before. What we can do here is nest the "Bottom 10 PC Customers" measure inside SUMX to force the values returned by this measure to be summed accordingly:

```
Bottom 10 PC Customers Summed =
SUMX(Customers,[Bottom 10 PC Customers])
```

As you can see in Figure 7-10, the Total row now calculates the total value for the bottom 10 percent of customers by total revenue.

CUSTOMER NAME	Bottom 10 PC Customers Summed
Charlottesville & Co	$77,190
Cheney & Co	$41,660
Kassel & Sons	$67,439
Morristown & Co	$96,208
Palo Alto Ltd	$95,801
Port Hammond Bros	$50,170
Rhodes Ltd	$88,086
Victoria Ltd	$93,788
Total	**$610,342**

Figure 7-10. *Summing the total revenue value for the "Bottom 10 PC Customers" measure*

To rank your customers or products, you can use the RANKX function to assign a ranking order, and you must also use this function to find the top or bottom percent. To analyze TopN entities, you can simply use a TopN visual level filter. See also QU63 "Can I Rank Values in Power Query".

Q49. How Do I Calculate New or Returning Customers in Any Month?

On the face it, this might seem a complex calculation if you're new to DAX. However, if we step back and break the analysis down into its constituent parts, it won't seem so daunting. The first consideration is how you intend to display these insights. The assumption is that the measures calculating the number of new and returning customers would be displayed in a table or matrix visual, with YEAR and MONTH columns from the DateTable categorizing the data, as shown in Figure 7-11.

YEAR	MONTH	New Customers	Returning Customers
2018	Jan	13	
	Feb	6	
	Mar	7	1
	Apr	5	5
	May	5	6
	Jun	5	3
	Jul	3	9
	Aug	4	11
	Sep	5	5
	Oct	2	4
	Nov	4	3

Figure 7-11. *You can analyze new or returning customers in a table or matrix using the YEAR and MONTH columns*

Let's initially take the "New Customers" measure and work through how this measure must be authored. The first step is to build a list of customers (in memory) with transactions in each of the months displayed in the visual, for example, customers with transactions in **March 2018**. To do this we can construct a DAX table expression where we use the VALUES function on the CUSTOMER ID field in the Sales table to generate a list of distinct customer IDs with sales in the relevant month. This would be the table expression:

```
CurrentCustomers = VALUES ( Sales[CUSTOMER ID] )
```

Note   You can't put table expressions into visualizations. The expressions shown here will be used as building blocks in the final measure later.

We must then build another in-memory table containing a list of customers who had transactions in the previous months to the month in question. Here we can use the CALCULATETABLE function that behaves just like CALCUATE but will return a table rather than a scalar value. CALCULATETABLE will filter the table generated by VALUES to hold the IDs of the customers with transactions that are before the earliest date in the relevant month. For example, in March 2018, this would be customers with transactions before **1 March 2018**:

```
PreviousCustomers =
        CALCULATETABLE (
            VALUES(Sales[CUSTOMER ID]),
                FILTER ( ALL ( DateTable ), DateTable[DATEKEY]
    < MIN ( DateTable[DATEKEY] ) ) )
```

Once we have the two in-memory tables generated by DAX, we can then calculate the number of customers who don't have IDs in both tables. In DAX we have a number of "set" functions that allow comparisons between tables to find matching and non-matching values. We can use the EXCEPT set function to generate a table of values containing the rows of our "CurrentCustomers" table that are not in the rows of the "PreviousCustomers" table.

In Figure 7-12, you can see the two tables generated by the DAX code for **March 2018**. We can see that CUSTOMER ID **78** is the only customer who had transactions in the previous months. Therefore, we can expect the "New Customers" measure to return **seven** new customers in March.

PreviousCustomers | CurrentCustomers

CUSTOMER ID		CUSTOMER ID
3		2
11		10
12		17
18		17
18		62
20		67
26		69
31		69
32		78
35		79
40		
52		
54		
61		
65		
68		
71		
72		
78		
82		

YEAR	MONTH	New Customers	Returning Customers
2018	Jan	13	
	Feb	6	
	Mar	7	1
	Apr	5	5
	May	5	6
	Jun	5	3
	Jul	3	9
	Aug	4	11
	Sep	5	5
	Oct	2	4
	Nov	4	3

Figure 7-12. *Generating in-memory tables to compare current customers with customers who have transactions in previous months*

Now we are ready to combine all the component parts of the calculation into this measure to calculate the number of new customers:

```
New Customers =
    VAR CurrentCustomers = VALUES ( Sales[CUSTOMER ID] )
    VAR PreviousCustomers =
        CALCULATETABLE (
            VALUES(Sales[CUSTOMER ID]),
                FILTER ( ALL ( DateTable ), DateTable[DATEKEY]
    < MIN ( DateTable[DATEKEY] ) ) )
    RETURN
    COUNTROWS ( EXCEPT ( CurrentCustomers, PreviousCustomers ) )
```

For returning customers, we can use the INTERSECT function that will return values from the "CurrentCustomers" table that are also in the "PreviousCustomers" table.

```
Returning Customers =
    VAR CurrentCustomers = VALUES ( Sales[CUSTOMER ID] )
    VAR PreviousCustomers =
        CALCULATETABLE (
            VALUES(Sales[CUSTOMER ID]),
                FILTER ( ALL ( DateTable ), DateTable[DATEKEY]
    < MIN ( DateTable[DATEKEY] ) ) )
    RETURN
    COUNTROWS ( INTERSECT ( CurrentCustomers, PreviousCustomers ) )
```

You can see both "New Customers" and "Returning Customers" in Figure 7-11. Understanding how these measures work means that we can make small edits to the code to calculate new or return customers for different time frames, for example, to calculate the number of customers in the current month that didn't have transactions in the previous month:

```
New Customers from Previous Mth =
    VAR CurrentCustomers = VALUES ( Sales[CUSTOMER ID] )
    VAR PreviousCustomers =
        CALCULATETABLE (
            VALUES(Sales[CUSTOMER ID]),
                PREVIOUSMONTH(DateTable[DATEKEY]))
    RETURN
    COUNTROWS ( EXCEPT ( CurrentCustomers, PreviousCustomers ) )
```

To find new and returning customers, we can harness the power of DAX to generate in-memory tables, and using set functions we can, therefore, compare values in such tables.

Q50. How Can I Compare Customers' Like-for-Like Sales?

Let's begin by exploring a situation where you might ask this question. For instance, your objective is to analyze the sales performance of your customers specifically for buying the product "Bordeaux" during the years 2021 and 2022. The comparison is only valid when comparing customers who bought "Bordeaux" in both years, not either year.

However, the challenge arises when using a slicer to select these two years. The slicer will filter customers based on their sales of "Bordeaux" in either of the chosen years, rather than considering sales only in *both* years. This issue is depicted in Figure 7-13.

WINE	CUSTOMER NAME	2021	2022	**Total**
■ Bordeaux	Barstow Ltd	$12,900	$25,800	**$38,700**
☐ Champagne	Brooklyn Ltd	$15,675		**$15,675**
☐ Chardonnay	Burningsuit Ltd	$22,950	$22,950	**$45,900**
☐ Chenin Blanc	Castle Rock Ltd	$31,650	$63,300	**$94,950**
	Chatou & Co	$25,650	$51,300	**$76,950**
☐ Chianti	Clifton Ltd	$33,375		**$33,375**
☐ Grenache	Issaquah & Co		$36,600	**$36,600**
	Leeds & Co		$9,750	**$9,750**
☐ Lambrusco	Littleton & Sons	$20,250	$40,500	**$60,750**
☐ Malbec	Melbourne Ltd	$21,000		**$21,000**
	New Castle Ltd		$23,550	**$23,550**
YEAR	Nizhny Novgorod & Co	$34,950	$56,250	**$91,200**
☐ 2018	Shanghai Ltd	$30,750		**$30,750**
☐ 2019	**Total**	**$249,150**	**$330,000**	**$579,150**
☐ 2020				
■ 2021				
■ 2022				
☐ 2023				

Figure 7-13. *Selecting years in the slicer returns customers with sales in either of the selected years, not both of the selected years*

Meanwhile, what we would like to see in the visual is shown in Figure 7-14.

WINE ⌄

■ Bordeaux

☐ Champagne

☐ Chardonnay

☐ Chenin Blanc

☐ Chianti

☐ Grenache

☐ Lambrusco

☐ Malbec

YEAR

☐ 2018

☐ 2019

☐ 2020

■ 2021

■ 2022

☐ 2023

CUSTOMER NAME	2021	2022	**Total**
Barstow Ltd	$12,900.00	$25,800.00	**$38,700.00**
Burningsuit Ltd	$22,950.00	$22,950.00	**$45,900.00**
Castle Rock Ltd	$31,650.00	$63,300.00	**$94,950.00**
Chatou & Co	$25,650.00	$51,300.00	**$76,950.00**
Littleton & Sons	$20,250.00	$40,500.00	**$60,750.00**
Nizhny Novgorod & Co	$34,950.00	$56,250.00	**$91,200.00**
Total	**$249,150.00**	**$330,000.00**	**$579,150.00**

Figure 7-14. *Selecting years in the slicer returns customers with sales in both the selected years*

You may remember in QU18 "How Can I Create an 'And' Slicer?" that we explored a similar problem. In answering QU18, we used the DISTINCTCOUNT function on the relevant column in the fact table (specifically the WINE ID column). However, in this example, we would have to pass the DISTINCTCOUNT function over the YEAR column, which does not reside in the fact table, only in the DateTable. Therefore, although we can take a similar approach, we need to work with columns in the DateTable.

To calculate which customers had sales in both years, we can generate an in-memory table using the YEAR column from the DateTable that will hold the years selected in the slicer. As part of this temporary in-memory table, we can calculate the "Total Revenue" value for the customer in the current filter. On the evaluations of "Barstow Ltd" and "Brooklyn Ltd," for example, the temporary table would hold the values as shown in Figure 7-15.

Barstow Ltd

YEAR	Total Revenue
2022	£25,800
2021	£12,900

Brooklyn Ltd

YEAR	Total Revenue
2021	£15,675

Figure 7-15. *Generating a temporary table to find the customers who have sales in the selected years*

You can now see that all we need to do is find the distinct values in the YEAR column of this temporary table using the following code:

```
Number of Years =
    CALCULATE (
        DISTINCTCOUNT ( DateTable[YEAR] ),
        FILTER ( ALLSELECTED ( DateTable[YEAR] ),
                             [Total Revenue]  ) )
```

This measure will return "2" for "Barstow Ltd" and "1" for "Brooklyn Ltd"; see Figure 7-16.

CUSTOMER NAME	2021	2022	**Total**
Barstow Ltd	2	2	**2**
Brooklyn Ltd	1	1	**1**
Burningsuit Ltd	2	2	**2**
Castle Rock Ltd	2	2	**2**
Chatou & Co	2	2	**2**
Clifton Ltd	1	1	**1**
Issaquah & Co	1	1	**1**
Leeds & Co	1	1	**1**
Littleton & Sons	2	2	**2**
Melbourne Ltd	1	1	**1**
New Castle Ltd	1	1	**1**
Nizhny Novgorod & Co	2	2	**2**
Shanghai Ltd	1	1	**1**
Total	**2**	**2**	**2**

Figure 7-16. *Showing the number of the selected years that each customer has sales*

Using the "Number of Years" code in a variable means that we can now construct a measure to filter those customers where this value equals the number of years selected in the slicer:

```
Like for Like Sales =
VAR NumberOfYears =
    CALCULATE (
     DISTINCTCOUNT ( DateTable[YEAR] ),
     FILTER ( ALLSELECTED ( DateTable[YEAR] ), [Total Revenue] ) )
VAR NumberSelectedYears =
    COUNTROWS ( ALLSELECTED ( 'DateTable'[YEAR] ) )
RETURN
    CALCULATE (
        [Total Revenue],
      FILTER ( Customers, NumberOfYears = NumberSelectedYears ) )
```

The results of this measure showing like-for-like sales are displayed in Figure 7-14.

Q51. How Do I Calculate Pareto or ABC Analysis?

The fact that your sales are primarily generated by repeat customers or that customers tend to buy their preferred products is one of those certainties of business life. It is worth noting, for example, that approximately 80 percent of your revenue would normally originate from just 20 percent of your customers. This phenomenon is commonly referred to as "Pareto" analysis or the "80/20" rule.

The "ABC" rule is a little more flexible. In this method of analysis, you can simply decide on the thresholds you are looking for. For example, you might consider that customers who contribute to up to 70% of the total sales are your best customers and therefore might be categorized as "A" category customers. Category "B" customers could be those contributing to the next 20% of sales, and category C customers produce the last 10% of sales. However, you can alter the thresholds as required. Let's explore how we can generate this method of analyzing our customers' sales.

We can start by categorizing our customers using calculated columns in the Customers dimension. In the first column we will calculate the total revenue for each customer:

```
Customer Revenue = [Total Revenue]
```

In the second column, for each customer, we must calculate the total revenue for customers that have a higher total revenue value. For example, "Castle Rock Ltd" has a total sales revenue of **$291,620**. There are two customers who have higher sales, "Concord Ltd" and "Issaquah & Co." Their combined total sales value is **$947,712**; see Figure 7-17.

CUSTOMER NAME	REGION ID	NO. OF STORES	Customer Revenue	Cumulated Revenue
Issaquah & Co	500	11	$338,880	$338,880
Concord Ltd	700	4	$317,212	$656,092
Castle Rock Ltd	1200	16	$291,620	$947,712
Ballard & Sons	1800	15	$264,262	$1,211,974
Fort Worth Ltd	200	19	$262,490	$1,474,464
Busan & Co	600	13	$262,410	$1,736,874
Germantown & Co	100	11	$260,441	$1,997,315
Burningsuit Ltd	1200	2	$257,055	$2,254,370

Figure 7-17. *Calculating the cumulative totals for customers with higher sales in a calculated column*

This is the code that will do this job:

```
Cumulated Revenue =
VAR CurrentRevenue = Customers[Customer Revenue]
VAR HighSalesCustomers =
 FILTER(Customers, Customers[Customer Revenue]>=CurrentRevenue)
RETURN
SUMX(HighSalesCustomers, Customers[Customer Revenue])
```

We must now calculate what percentage of the cumulative values is of the grand total sales:

```
Cumulated PC =
DIVIDE(Customers[Cumulated Revenue],SUM(Sales[REVENUE]))
```

In Figure 7-18, we can see the "Castle Rock Ltd" sales revenue along with the higher sales constitute **6.92%** of the total sales.

CUSTOMER NAME	REGION ID	NO. OF STORES	Customer Revenue	Cumulated Revenue	Cumulated PC
Issaquah & Co	500	11	$338,880	$338,880	2.47%
Concord Ltd	700	4	$317,212	$656,092	4.79%
Castle Rock Ltd	1200	16	$291,620	$947,712	6.92%
Ballard & Sons	1800	15	$264,262	$1,211,974	8.85%
Fort Worth Ltd	200	19	$262,490	$1,474,464	10.76%
Busan & Co	600	13	$262,410	$1,736,874	12.68%
Germantown & Co	100	11	$260,441	$1,997,315	14.58%
Burningsuit Ltd	1200	2	$257,055	$2,254,370	16.46%
Branch Ltd	300	23	$253,026	$2,507,396	18.30%
Liverpool & Sons	1000	22	$249,550	$2,756,946	20.13%
Clifton Ltd	1100	15	$246,915	$3,003,861	21.93%
Landstuhl Ltd	1800	21	$243,246	$3,247,107	23.70%
Lavender Bay Ltd	500	10	$241,784	$3,488,891	25.47%

Figure 7-18. *Calculating the percentage cumulative total of the grand total sales*

Finally, we can categorize these percentages into "A," "B," or "C" as required:

```
ABC =
    SWITCH (
        TRUE,
        customers[Cumulated PC] <= 0.7, "A",
        customers[Cumulated PC] <= 0.9, "B",
        "C"
        )
```

We can see the calculated columns that generate the "ABC" analysis in Figure 7-19 and that "Castle Rock Ltd" is in category "A."

CUSTOMER NAME	REGION ID	NO. OF STORES	Customer Revenue	Cumulated Revenue	Cumulated PC	ABC
Issaquah & Co	500	11	$338,880	$338,880	2.47%	A
Concord Ltd	700	4	$317,212	$656,092	4.79%	A
Castle Rock Ltd	1200	16	$291,620	$947,712	6.92%	A
Ballard & Sons	1800	15	$264,262	$1,211,974	8.85%	A
Fort Worth Ltd	200	19	$262,490	$1,474,464	10.76%	A
Busan & Co	600	13	$262,410	$1,736,874	12.68%	A
Germantown & Co	100	11	$260,441	$1,997,315	14.58%	A
Burningsuit Ltd	1200	2	$257,055	$2,254,370	16.46%	A
Branch Ltd	300	23	$253,026	$2,507,396	18.30%	A
Liverpool & Sons	1000	22	$249,550	$2,756,946	20.13%	A
Clifton Ltd	1100	15	$246,915	$3,003,861	21.93%	A
Landstuhl Ltd	1800	21	$243,246	$3,247,107	23.70%	A
Lavender Bay Ltd	500	10	$241,784	$3,488,891	25.47%	A
Jacksonville Ltd	100	1	$239,939	$3,728,830	27.22%	A
Melbourne Ltd	100	11	$236,864	$3,965,694	28.95%	A

Figure 7-19. *Generating the "ABC" analysis we can see that "Castle Rock Ltd" is in category "A"*

We can now use these categories to perform analysis on our customers. We could, for example, generate a matrix that calculated subtotals for each category; see Figure 7-20.

ABC	CUSTOMER NAME	Total Revenue
A	Kennebunkport & Co	$178,737
	Landstuhl Ltd	$243,246
	Loveland & Co	$187,655
	Villeneuve-d'Ascq Ltd	$172,100
	Total	**$781,738**
B	Shanghai Ltd	$134,832
	Spokane Ltd	$115,797
	Total	**$250,629**
C	Waterbury Ltd	$107,544
	Yokohama & Co	$104,395
	Total	**$211,939**
	Total	**$1,244,306**

Figure 7-20. *Analyzing "ABC" categories in a matrix (we are only showing eight customers here)*

However, the downside of generating calculated columns to categorize customers is that we can't browse sales revenue by other entities. We may, for instance, need to show "ABC" analysis against specific product sales or in specific sales regions. For this, we must author this measure that follows the same logic as the calculated columns to generate each category:

```
ABC Customers =
VAR Revenue =
    CALCULATE ( [Total Revenue] )
VAR SumRevenue =
    CALCULATE ( [Total Revenue], ALL ( Customers ) )
VAR CumulativeRevenue =
    CALCULATE (
        [Total Revenue],
        FILTER ( ALL ( Customers ), [Total Revenue] >= Revenue ) )
VAR ABCPercent =
        CumulativeRevenue / SumRevenue
RETURN
    SWITCH (
        TRUE (),
        ABCPercent <= 0.7, "A",
        ABCPercent <= 0.9, "B",
        "C")
```

We can now put this measure into a matrix visual and using a visual level filter analyze sales for "A" category customers for "Bordeaux" sales; see Figure 7-21.

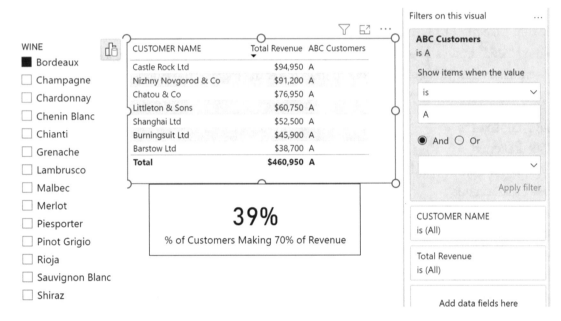

Figure 7-21. *Using the "ABC Customers" measure filtering category "A" customers for product "Bordeaux"*

Notice in Figure 7-21 we have an additional measure displayed in the card visual:

```
% of Customers Making 70% of Revenue =
[No. of Customers Making 70% of Revenue] /
DISTINCTCOUNT ( Sales[CUSTOMER ID] )
```

Hopefully, you will find that by applying the "ABC" rule you can find those customers or products that are most profitable to your sales revenue or indeed make strategies to engage the less lucrative customers or promote the less lucrative products.

CHAPTER 8

Employees

It would appear that two of the most important metrics when analyzing your employees are what they cost your organization and the frequency with which they quit. In this section, we look at how we can uncover this information by answering just two questions: How do you calculate changing payments over time, and how can you analyze employee attrition and churn rate?

Q52. How Do I Calculate Employees Changing Salary Over Time?

Let's start by focusing on the source data for this question. In the following scenario, our salespeople are our employees. Our salespeople get paid daily with different daily rates each year, as shown in the "Salary" table in Figure 8-1.

SALESPERSON ID	From	To	Daily
1	01 January 2016	31 December 2016	$141
2	01 January 2016	31 December 2016	$127
3	01 January 2016	31 December 2016	$177
4	01 January 2016	31 December 2016	$123
5	01 January 2016	31 December 2016	$152
6	01 January 2016	31 December 2016	$160
1	01 January 2017	31 December 2017	$160
2	01 January 2017	31 December 2017	$154
3	01 January 2017	31 December 2017	$171
4	01 January 2017	31 December 2017	$182
5	01 January 2017	31 December 2017	$148
6	01 January 2017	31 December 2017	$155
1	01 January 2018	31 December 2018	$132
2	01 January 2018	31 December 2018	$141
3	01 January 2018	31 December 2018	$158

Figure 8-1. *The "Salary" table*

© Alison Box 2023
A. Box, *A Power BI Compendium*, https://doi.org/10.1007/978-1-4842-9765-0_8

The "Sales" table holds daily sales transactions for each salesperson; see Figure 8-2.

SALEDATE	SALESPERSON ID	SALES
01 January 2017	6	$96
01 January 2017	3	$184
01 January 2017	1	$185
02 January 2017	4	$131
03 January 2017	1	$195
06 January 2017	6	$148
06 January 2017	5	$167
07 January 2017	6	$193
07 January 2017	2	$238
07 January 2017	3	$239
08 January 2017	3	$217
09 January 2017	6	$154
10 January 2017	4	$181
10 January 2017	2	$92

Figure 8-2. *The "Sales" table*

In our data model, both the Salary table and the Sales table are related to the SalesPeople dimension in a many-to-one relationship. Each salesperson has multiple daily payment rates and multiple sales transactions.

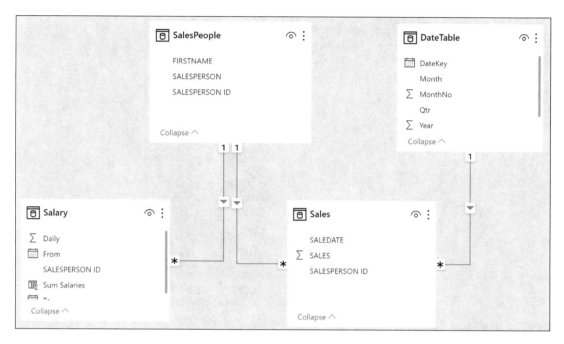

Figure 8-3. *The data model showing the many-to-one relationships between the tables*

We can now write a very simple measure that sums the daily salaries:

```
Total Salary = SUM ( Salary[Daily] )
```

However, the more challenging task is to calculate the monthly salary paid to each salesperson, for example, salesperson "Abel," as shown in Figure 8-4.

SALESPERSON ▲	Year	Month	Salary Paid
⊟ Abel	⊟ 2017	Jan	$1,760
		Feb	$1,440
		Mar	$320
		Apr	$1,120
		May	$1,600
		Jun	$640
		Jul	$1,120
		Aug	$1,120
		Sep	$1,600
		Oct	$1,440
		Nov	$1,440
		Dec	$1,120
		Total	**$14,720**

Figure 8-4. Salesperson, "Abel's" monthly salary

So that we know the salary that was paid for each day in the Sales table, the Salary table records date ranges into which the transaction dates must fall. For example, salesperson "Abel" will be paid **$160** for every day in 2017 that has transactions recorded against them.

We can begin calculating salaries paid by authoring a calculated column in the Salary table. The calculation we need must use a temporary sales table that groups transactions by date and salesperson. We can do this using SUMMARIZE. Then we can filter this temporary table according to each salesperson and the date range recorded in the Salary table, calculating the "Total Salary" measure for the filtered dates:

```
Sum Salaries =
SUMX (
    FILTER (
        SUMMARIZE(Sales, Sales[SALEDATE], Sales[SALESPERSON ID]),
        Sales[SALEDATE] >= Salary[From]
            && Sales[SALEDATE] <= Salary[To]
            && Sales[SALESPERSON ID] = Salary[SALESPERSON ID]
    ),
    [Total Salary]
)
```

In Figure 8-5, you can see that Salesperson ID **1** should be paid **$14,720** in 2017 based on **$160** per day. This is correct as this salesperson worked 92 days this year.

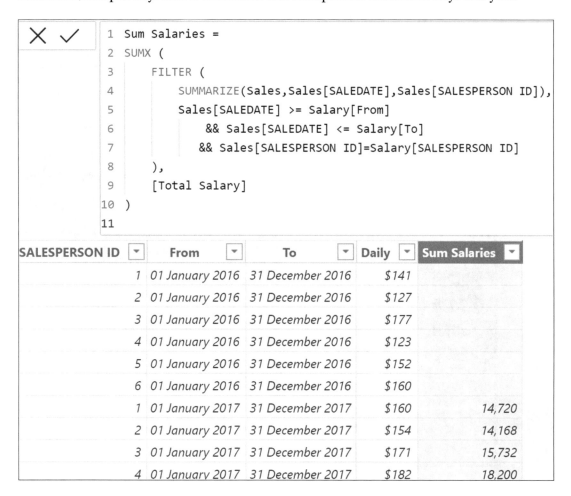

```
 1  Sum Salaries =
 2  SUMX (
 3      FILTER (
 4          SUMMARIZE(Sales,Sales[SALEDATE],Sales[SALESPERSON ID]),
 5          Sales[SALEDATE] >= Salary[From]
 6              && Sales[SALEDATE] <= Salary[To]
 7              && Sales[SALESPERSON ID]=Salary[SALESPERSON ID]
 8      ),
 9      [Total Salary]
10  )
11
```

SALESPERSON ID	From	To	Daily	Sum Salaries
1	01 January 2016	31 December 2016	$141	
2	01 January 2016	31 December 2016	$127	
3	01 January 2016	31 December 2016	$177	
4	01 January 2016	31 December 2016	$123	
5	01 January 2016	31 December 2016	$152	
6	01 January 2016	31 December 2016	$160	
1	01 January 2017	31 December 2017	$160	14,720
2	01 January 2017	31 December 2017	$154	14,168
3	01 January 2017	31 December 2017	$171	15,732
4	01 January 2017	31 December 2017	$182	18,200

Figure 8-5. *Calculating the "Total Salary" value in the context of a calculated column in the Salary table*

However, if we want to browse the salaries by salespeople, month, or year, we must generate a measure. Now that we know the calculated column returns the correct value, to create a measure, we can simply wrap the entire expression inside SUMX, remembering that it's the Salary table that we must iterate:

```
Salary Paid =
SUMX (
    Salary,
SUMX (
```

```
FILTER (
    SUMMARIZE(Sales, Sales[SALEDATE],Sales[SALESPERSON ID]),
    Sales[SALEDATE] >= Salary[From]
        && Sales[SALEDATE] <= Salary[To]
        && Sales[SALESPERSON ID]=Salary[SALESPERSON ID]
    ),
    [Total Salary]
)
)
```

We can now put this measure into a visual and browse our salespeople's salaries by month or year using the respective columns from the DateTable; see Figure 8-6.

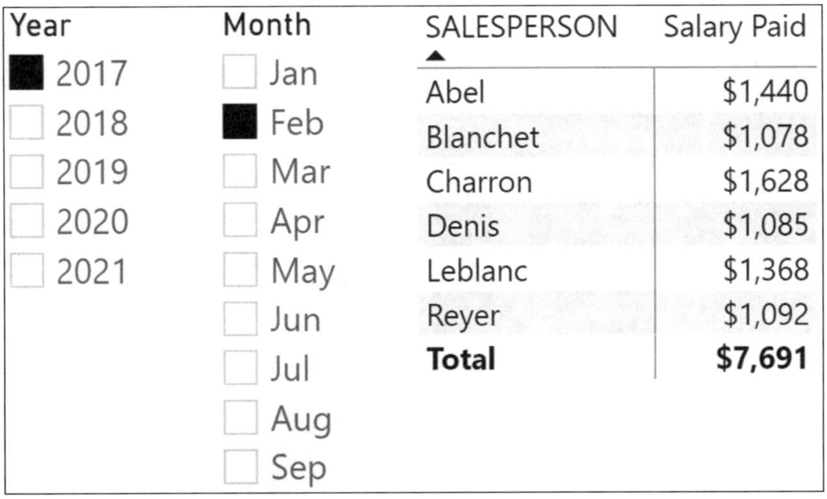

Year	Month	SALESPERSON ▲	Salary Paid
■ 2017	☐ Jan	Abel	$1,440
☐ 2018	■ Feb	Blanchet	$1,078
☐ 2019	☐ Mar	Charron	$1,628
☐ 2020	☐ Apr	Denis	$1,085
☐ 2021	☐ May	Leblanc	$1,368
	☐ Jun	Reyer	$1,092
	☐ Jul	**Total**	**$7,691**
	☐ Aug		
	☐ Sep		

Figure 8-6. *Browsing salespeople's monthly salaries*

So again, what appears to be a challenging calculation becomes less so if it's taken one step at a time.

Q53. How Do I Calculate Employee Turnover in Each Month?

For this, we must have a correctly structured data model that includes a People dimension, Figure 8-7.

EMPNO	DOB	GENDER	ETHNICITY	Age	Age Band
1000	30 October 1989	M	Black, Black British, Caribbean or African	33	31 to 40
1001	22 January 1970	F	Black, Black British, Caribbean or African	53	51 to 60
1002	09 March 1954	F	Black, Black British, Caribbean or African	69	61 and Greater
1003	08 August 1988	F	White	34	31 to 40
1004	10 July 1963	M	White	59	51 to 60
1005	09 December 1970	M	Black, Black British, Caribbean or African	52	51 to 60
1006	16 December 1991	F	Any other White background	31	31 to 40
1007	14 January 1978	F	Asian or Asian British	45	41 to 50
1008	14 August 1958	M	White	64	61 and Greater
1009	24 November 1996	M	Any other White background	26	21 to 30
1010	30 June 1987	F	Mixed or multiple ethnic groups	35	31 to 40
1011	16 June 1952	F	Any other White background	70	61 and Greater
1012	16 October 1962	F	Asian or Asian British	60	51 to 60
1013	23 December 1959	M	White	63	61 and Greater
1014	26 July 1955	F	Any other White background	67	61 and Greater
1015	16 March 1994	F	White	29	21 to 30
1016	04 March 2003	F	White	20	Under 21
1017	21 October 1999	F	White	23	21 to 30
1018	08 August 1969	M	Mixed or multiple ethnic groups	53	51 to 60
1019	28 August 1956	F	Any other White background	66	61 and Greater
1020	17 December 1973	F	Mixed or multiple ethnic groups	49	41 to 50
1021	04 October 1999	M	White	23	21 to 30

Figure 8-7. *The People dimension*

We must then have an Employment fact table that records the hire and leave dates of each employee; see Figure 8-8.

EMPNO	HIRE DATE	LEAVE DATE	ROLE
1000	25 August 2018	09 February 2021	Technician
1001	20 May 2022		Administrator
1001	25 October 2020	20 May 2022	Team Leader
1002	30 March 2019		Team Leader
1003	20 September 2018	06 October 2021	Administrator
1004	15 July 2022		Technician
1004	08 April 2021	09 October 2021	Technician
1004	09 October 2021	15 July 2022	Administrator
1005	23 November 2021		Technician
1005	27 October 2020	23 November 2021	Intern
1005	03 January 2019	27 October 2020	Administrator
1006	30 January 2022		Manager
1007	02 April 2022		Team Leader
1008	11 August 2021		Manager
1008	01 September 2020	11 August 2021	Technician
1009	28 December 2020	01 January 2021	Intern

Figure 8-8. *The Employment fact table*

And of course, we must have a date dimension that can filter dates according to both hire and leave dates. Therefore, the date dimension must be related to the Employment table using the DateKey column that links to ***both*** the HIRE DATE and the LEAVE DATE columns, the latter relationship being set to inactive; see Figure 8-9.

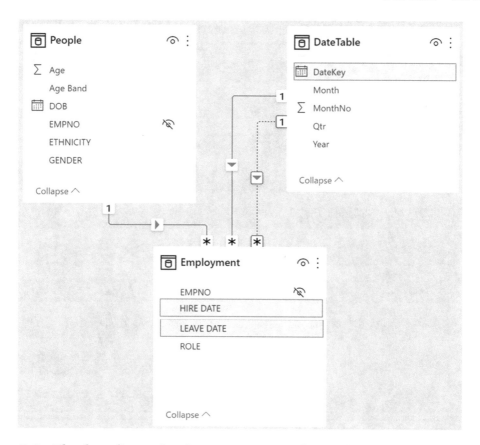

Figure 8-9. *The date dimension has an inactive relationship between the DateKey column and the LEAVE DATE column*

Now that we have our data model in place, we can start to author the measures we require to calculate monthly staff turnover. These will be as follows:

- **Head Count** – the number of current employees each month

- **No. of Joiners** – people joining or re-joining the organization

- **No. of New People** – the number of joiners who have not re-joined

- **No. of Leavers** – the number of people who have left the company

- **No. of Leavers To Date** – monthly cumulative leavers

You can see these measures displayed in a table visual in Figure 8-10.

Year	Head Count	No. of Joiners	No. of New People	No. of Leavers	No. of Leavers To Date
⊟ **2018**	**91**	**102**	**97**	**11**	**11**
Jan	9	9	9		
Feb	18	9	9		
Mar	25	9	8	2	2
Apr	36	11	11		2
May	44	8	8		2
Jun	48	5	5	1	3
Jul	53	5	5		3
Aug	60	10	9	3	6
Sep	73	14	14	1	7
Oct	82	9	9		7
Nov	86	6	5	2	9
Dec	91	7	5	2	11
⊟ **2019**	**163**	**108**	**98**	**36**	**47**
Jan	98	9	9	2	13
Feb	103	8	7	3	16
Mar	116	16	15	3	19
Apr	120	7	7	3	22
May	126	7	7	1	23
Jun	131	8	8	3	26

Figure 8-10. *A table visual showing monthly headcount, joiners, and leavers*

This is the DAX code you will need for each of these measures:

```
Head Count =
COUNTROWS (
    FILTER (
        ALLSELECTED ( Employment ),
        Employment[HIRE DATE] <= MAX ( DateTable[Datekey] )
            && Employment[LEAVE DATE] > MAX ( DateTable[Datekey] )
            || Employment[HIRE DATE] <= MAX ( DateTable[Datekey] )
                && ISBLANK ( Employment[LEAVE DATE] )
    )
)
```

In this measure, the FILTER function iterates a virtual table built by the ALLSELECTED function containing all the rows from the Employment but retaining any filters coming through from slicers on the canvas, for example, for slicing by Gender or Age Band. It then takes the last date of the month in the current filter, for example, if we are analyzing March 2018, this would be 31 March. Then the expression counts the number of employees who were hired before or on this date and have either left after this date or not left.

```
No. of Joiners =
COUNTROWS ( Employment )
```

The "No. of Joiners" measure is a simple count of the rows in the Employment table in the current filter. For example, if March 2018 is in the current filter, the Employment table will be filtered for all employees who started this month, using the active relationship between the DateTable and the Employment fact table.

```
No. of New People =
    VAR CurrentEmps = VALUES ( Employment[EMPNO] )
    VAR PreviousEmps =
        CALCULATETABLE (VALUES(Employment[EMPNO]),
                FILTER ( ALL ( DateTable ), DateTable[DATEKEY]
    < MIN ( DateTable[DATEKEY] ) ) )
    RETURN
    COUNTROWS ( EXCEPT ( CurrentEmps, PreviousEmps ) )
```

The "No. of New People" measure follows the same logic as the "New Customers" measure we explored in Q49 in that it builds two virtual tables whose rows are then compared.

The "CurrentEmps" variable uses the VALUES function to generate a virtual table of employees in the Employment table in the current filter, for example, current employees in March 2018.

The "PreviousEmps" variable uses the FILTER function to generate a virtual DateTable that contains rows where the date is earlier than the date in the current filter, for example, before 1 March. The CALCULATETABLE function then uses the table generated by the FILTER function to filter employees in the VALUES table to hold only employees who started before 1 March.

The EXCEPT function generates a table containing all the rows in the first variable (e.g., the current employees in March 2018) that are not in the second variable (i.e., employees in months prior to March 2018) and the rows are counted.

To calculate the number of leavers, we must use the inactive relationship between the DateTable and the Employment table so that the LEAVE DATE will be filtered by the current filter in the table visual:

```
No. of Leavers =
CALCULATE (
    COUNTROWS ( Employment ),
    USERELATIONSHIP ( Employment[LEAVE DATE], DateTable[DateKey] ) )
```

Now let's see how many leavers we have to date:

```
No. of Leavers To Date =
  COUNTROWS (
        FILTER (
            ALLSELECTED ( Employment ),
            Employment[LEAVE DATE]
                <= MAX ( DateTable[DateKey] )
                && NOT ( ISBLANK ( Employment[LEAVE DATE] ) )
) )
```

The "No. of Leavers to Date" measure removes the filter from the Employment table using the ALLSELECTED function. It then calculates the number of rows in the Employment table whose LEAVE DATE is less than or equal to the latest date in the current filter context and is not blank.

The problem, however, with both the "Head Count" and "No. Leavers To Date" measures is that, because we have not removed any filters coming through from the date table, they will respond to a filter on the YEAR column. Therefore, if a year is selected in a slicer, these measures will show only values for the selected year and not a cumulative value. Consider Figure 8-11 where we have two versions of a table visual showing both these measures. On the left, the data is not filtered by the Year slicer, and we can see that, for 2019, the head count is **163**. However, when we filter by the Year slicer, the table on the right shows an incorrect headcount value for 2019 of **94.**

Not Filtered			Filtered				
Year ▲	Head Count	No. of Leavers To Date	Year ▲	Head Count	No. of Leavers To Date	Year	
⊟ 2019	**163**	**47**	⊟ 2019	94	14	☐ 2018	
Jan	98	13	Jan	9		■ 2019	
Feb	103	16	Feb	17		☐ 2020	
Mar	116	19	Mar	31	2	☐ 2021	
Apr	120	22	Apr	37	3	☐ 2022	
May	126	23	May	44	3		
Jun	131	26	Jun	50	5		
Jul	143	28	Jul	64	5		
Aug	144	32	Aug	67	7		
Sep	151	36	Sep	75	10		
Oct	153	37	Oct	78	10		
Nov	158	41	Nov	84	13		
Dec	163	47	Dec	94	14		
⊟ 2020	220	96	**Total**	94	14		
Jan	171	48					
Feb	170	52					
Mar	178	58					
Total	**297**	**269**					

Figure 8-11. *The "Head Count" and "No. of Leavers To Date" measures do not calculate correctly with a filter from the DateTable*

To remedy this, we need to write more complex code that will remove the filter from the DateTable and then refilter the DateTable for years up to the year in the current filter context. This is the code we need for each of the measures:

```
Head Count =
    CALCULATE (
        COUNTROWS (
            FILTER (
                ALLSELECTED ( Employment ),
                Employment[HIRE DATE] <= MAX ( DateTable[Datekey] )
                    && Employment[LEAVE DATE] > MAX ( DateTable[Datekey] )
                    || Employment[HIRE DATE] <= MAX ( DateTable[Datekey] )
                        && ISBLANK ( Employment[LEAVE DATE] )
            )
        ),
        FILTER ( ALL ( DateTable ), DateTable[DateKey] <= MAX (
        DateTable[DateKey] ) )
    )
```

```
No. of Leavers To Date =
    CALCULATE (
  COUNTROWS (
        FILTER (
            ALLSELECTED ( Employment ),
            Employment[LEAVE DATE]
                <= MAX ( DateTable[DateKey] )
                && NOT ( ISBLANK ( Employment[LEAVE DATE] ) )

) ),
        FILTER ( ALL ( DateTable ), DateTable[DateKey] <= MAX (
        DateTable[DateKey] ) )
    )
```

We can now see in Figure 8-12 that the measures calculate correctly when we filter by year.

	Not Filtered			Filtered			
Year	Head Count	No. of Leavers To Date	Year	Head Count	No. of Leavers To Date		Year
⊟ 2019	163	47	⊟ 2019	163	47		☐ 2018
Jan	98	13	Jan	98	13		■ 2019
Feb	103	16	Feb	103	16		☐ 2020
Mar	116	19	Mar	116	19		
Apr	120	22	Apr	120	22		☐ 2021
May	126	23	May	126	23		☐ 2022
Jun	131	26	Jun	131	26		
Jul	143	28	Jul	143	28		
Aug	144	32	Aug	144	32		
Sep	151	36	Sep	151	36		
Oct	153	37	Oct	153	37		
Nov	158	41	Nov	158	41		
Dec	163	47	Dec	163	47		
⊟ 2020	220	96	Total	163	47		
Jan	171	48					
Feb	170	52					
Mar	178	58					
Total							

Figure 8-12. *The "Head Count" and "No. of Leavers To Date" measures are now correct*

Let's now compare the monthly headcount with the previous year. In Figure 8-13, we are again slicing by the Year column in the DateTable and comparing months in **2019** to the previous year's months. We are visualizing these head count values in a table visual and a line chart.

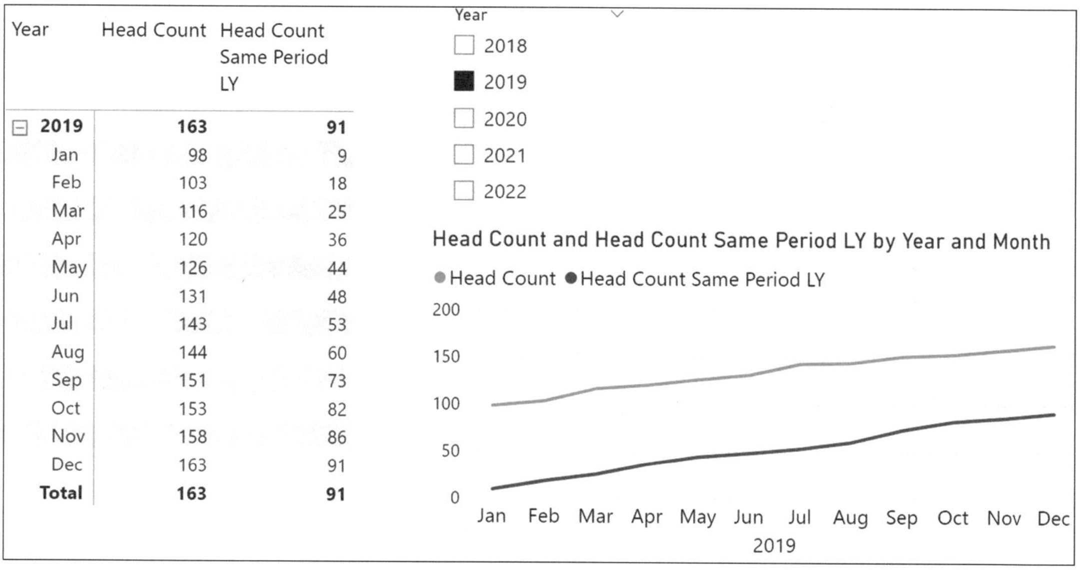

Year	Head Count	Head Count Same Period LY
□ **2019**	**163**	**91**
Jan	98	9
Feb	103	18
Mar	116	25
Apr	120	36
May	126	44
Jun	131	48
Jul	143	53
Aug	144	60
Sep	151	73
Oct	153	82
Nov	158	86
Dec	163	91
Total	**163**	**91**

Figure 8-13. *The "Head Count" and "Head Count Same Period LY" measures*

The "Head Count Same Period LY" measure nests the head count calculation inside CALCULATE to evaluate this using the SAMEPERIODLASTYEAR time intelligence function:

```
Head Count Same Period LY =
  CALCULATE ([Head Count], SAMEPERIODLASTYEAR ( DateTable[DateKey] )
```

We can now begin to consider the analysis of the various categories of employees. We'll take the age band category as an example using the "Age Band" column in the People dimension. Calculations for other categories, such as gender and ethnicity, would be the same, just substituting the relevant category column in the DAX code. In Figure 8-14, we are calculating head count for the different age bands for a selected year and the previous year.

Age Band	Age Band Head Count	Age Band Previous Yr	Year
Under 21	18	16	☐ 2018
21 to 30	40	37	☐ 2019
31 to 40	46	39	☐ 2020
41 to 50	46	40	☑ 2021
51 to 60	43	36	☐ 2022
61 and Greater	70	52	
Total	**263**	**220**	

Figure 8-14. *Analyzing head count for age bands across years*

These are the measures used in Figure 8-14:

```
Age Band Head Count =
    CALCULATE (
        COUNTROWS ( Employment ),
        FILTER (
            ALL ( Employment ),
            Employment[HIRE DATE]
< MAX ( DateTable[Datekey] )
                && Employment[LEAVE DATE]
>= MAX ( DateTable[Datekey] )
                || Employment[HIRE DATE]
< MAX ( DateTable[Datekey] )
                && ISBLANK ( Employment[LEAVE DATE] )
        ),
        VALUES ( People[Age Band] ) )

Age Band Previous Yr =
CALCULATE ( [Age Band Head Count], PREVIOUSYEAR(DateTable[DateKey]))
```

In the "Age Band Head Count" measure, note the use of the VALUES function to reapply the age band filter coming from the table visual that was lost by using the ALL function over the Employment table.

This chapter has been particularly pedestrian, taking you through copious lines of DAX code to arrive at the calculations required for the employee analysis. However, it's the hope that you can repurpose this code or indeed edit and expand on it as your requirements see fit.

CHAPTER 9

Power Query

People new to Power BI often approach Power Query in the same way they approach data modeling, that is, by ignoring it. This is probably because unless you specifically ask to transform your data, it will be imported into Power BI Desktop relatively unchanged. People often don't realize that whenever they load or refresh their data, it will pass through the Power Query funnel.

However, just like understanding the importance of a robust data model, Power Query is instrumental in shaping the data to achieve this goal. It would be a very unusual event indeed if all the tables you imported for your report fell neatly into a star schema! It's in Power Query that we split and merge tables to generate dimensions and fact tables and, of course, filter data not required by the report. Power Query is, on the whole, a very intuitive tool to use; filtering columns and rows, splitting columns and replacing values, for instance, can all be done from buttons on the Home tab of the Power Query window. All that most transformations require you to know is on which tab the relevant button sits.

In this last chapter, therefore, we concentrate on answers that require transformations that are not so intuitive or straightforward to achieve. One word of warning, however, Power Query uses a query language called "M," and in the examples that follow often you must enter or edit the M code. Unlike DAX, M is case-sensitive so please ensure that you type the code exactly as shown. References to column names must be exactly in the case they appear in the data.

To open Power Query, in Power BI desktop, use the "**Transform Data**" button on the **Home** tab.

Q54. Why Does My Query Fail on Refresh When I Rename Source Columns?

The most common reason why queries fail on refresh is because columns in the source data have been removed or renamed. However, this only happens when connecting to unstructured data sources, such as Excel or CSV files. The reason for this is that Power Query employs automatic data type detection when dealing with this type of data. Consequently, on import, two steps will automatically be added in the APPLIED STEPS pane as follows:

- **Promote Headers** – In this step, values in the first row of the table will become the column headers. However, this step is not applied if connecting to an Excel table where the column headers have already been defined.

- **Changed Type** – This step converts the values in each column from the "Any" data type to a specific data type based on Power Query's inspection of the data in each column.

To take a closer look at this, we've imported data from an Excel table and selected the "Changed Type" step in the APPLIED STEPS pane. In Figure 9-1 you can see in the Formula bar that the M code references each source column name.

Note   To turn on the Formula bar, use the **View** tab.

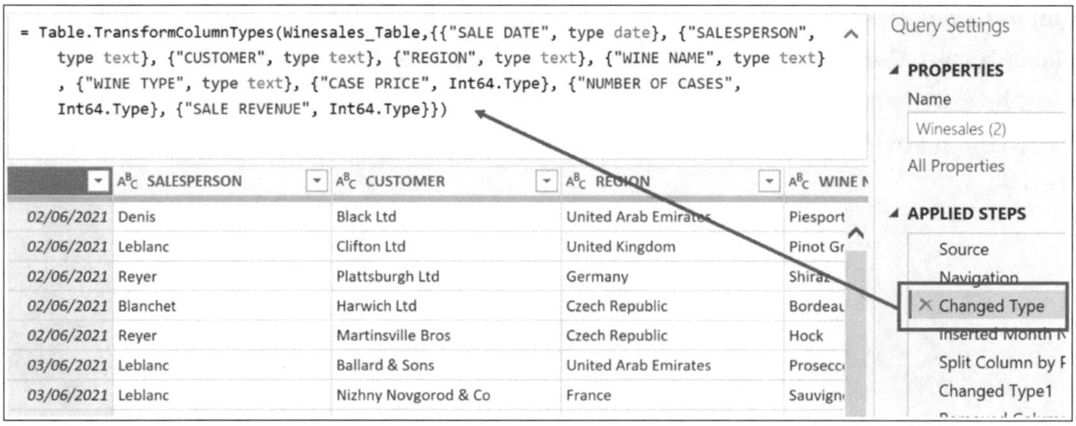

Figure 9-1. *The "Changed Type" step in the APPLIED STEPS pane*

Therefore, if we remove or rename any of these columns, for example, rename SALE DATE to SALEDATE, this step will fail on refresh and produce an error as shown in Figure 9-2.

Figure 9-2. *This error shows if you rename or remove source columns*

In fact, we don't necessarily need this step at all! The only columns that must have a data type assigned to them are date and numeric columns to facilitate calculations in Power BI Desktop. Therefore, there are several strategies we can adopt to avoid this problem.

We could edit the M code. The error message will identify the miscreant column or columns, and we can simply correct the names.

We could remove the "Changed Type" step. Then as the last step in the query, we can select all the columns and, using the **Detect Data Type** button on the **Transform** tab, assign data types to these columns. By making the "Changed Type" step last, if this step fails, at least all the previous steps will work.

We can stop Power Query from generating the "Changed Type" step. To do this we must disable automatic data type detection that can be done globally or just for the current report. In Power BI Desktop, on the **File** tab, under **Options and settings** and **Options**, we can change the setting as shown in Figure 9-3.

GLOBAL	Type Detection
Data Load	○ Always detect column types and headers for unstructured sources
Power Query Editor	○ Detect column types and headers for unstructured sources according to each file's setting
DirectQuery	◉ Never detect column types and headers for unstructured sources
CURRENT FILE	Type Detection
Data Load	☐ Detect column types and headers for unstructured sources
Regional Settings	
Privacy	

Figure 9-3. *You can turn off automatic data type detection*

Of course, you could just ensure that if connecting to Excel or CSV data, no one edits the column names or deletes the columns in these files.

Q55. Can I Import a Word Document into Power BI?

Yes, you can! However, you won't find the option in the **Get Data** list that is accessed from the **Get Data** button on the **Home** tab of Power BI Desktop. This is because Power BI can't import Word documents. What it can do, though, is to import text formatted as html with the help of a custom visual, and we can, of course, save Word documents as html. What we can then do is use Power Query to build a one-column table containing all the text in the html file that can then be rendered by the custom visual.

If the objective is to annotate your report, you can simply use a text box, but here the text formatting options are limited. You may want to include a table or different numbering styles, for example. Also, it makes sense that if you require lengthy annotations, it would be easier to type them up in Word. For these reasons, you may have a requirement to import Word documents.

Let's first create a document using Word, part of which you can see in Figure 9-4.

Heading 1

Video provides a powerful way to help you prove your point. When you click Online
You can also type a keyword to search online for the video that best fits your docume
header, footer, cover page, and text box designs that complement each other. For exa

Heading 2

Click Insert and then choose the elements you want from the different galleries. The
click Design and choose a new Theme, the pictures, charts, and SmartArt graphics ch
change to match the new theme. Save time in Word with new buttons that show up

Heading 3

1. To change the way a picture fits in your document,
2. click it and a button for layout options appears next to it.
3. When you work on a table,
4. click where you want to add a row or a column, and then click the plus sign.

Heading 4

- To change the way a picture fits in your document,
- click it and a button for layout options appears next to it.
- When you work on a table,

Jan	Feb	Mar
1,2345	3,9756	3,1065
3,9876	2,0234	2,5691
2,8412	5,0234	1,7185
1,9275	3,1745	1,9386

Figure 9-4. *Part of a Word document*

This document must be saved as an html file. Use **Save As** and select "Web Page,
Filtered (*.htm,*.html)" as the file type. This will remove proprietary Microsoft tags and
attributes. We've called our document "Word Document"; see Figure 9-5.

File name:	Word Document.htm
Save as type:	Web Page, Filtered (*.htm;*.html)
Authors:	Word Document (*.docx)
	Word Macro-Enabled Document (*.docm)
	Word 97-2003 Document (*.doc)
Page	Word Template (*.dotx)
	Word Macro-Enabled Template (*.dotm)
	Word 97-2003 Template (*.dot)
	PDF (*.pdf)
	XPS Document (*.xps)
	Single File Web Page (*.mht;*.mhtml)
le Folders	Web Page (*.htm;*.html)
	Web Page, Filtered (*.htm;*.html)

Figure 9-5. *Save your document as "Web Page, Filtered"*

Now we're ready to import this html file into a Power BI report. However, to render the html file, we must use a custom visual. We will use the custom visual, "HTML Content." To install this visual, in Power BI desktop, from the Visualizations gallery, select "**Get more visuals**." Then in the AppSource window, search for the "HTML Content" visual, and use the **Add** button to download it; see Figure 9-6.

Figure 9-6. *Adding the "HTML Content" custom visual*

In Power Query, on the dropdown of the **New Source** button, select **Blank Query**, and on the **View** tab, click on the **Advanced Editor** button. We now have a blank canvas on which to build our query.

What we will do here is build a single-column table whose single row contains all the content of our html file. This is the M code you can copy and paste into the Advanced Editor, changing the file path as required; see Figure 9-7.

```
let
Source = Text.FromBinary(File.Contents
    ("C:\Users\UserName\Desktop\Word Document.htm"),
    TextEncoding.Windows),
    #"Convert to table" =
    Table.FromRecords({[Imported Text = Source]})
in
#"Convert to table"
```

Figure 9-7. *Adding the M code to the Advanced Editor canvas*

This M code takes our file, converts it to a text file, and places it into a one-column, one-row table using the Table.FromRecords function. Inside the square brackets of this function, whatever you use on the left of "Source" becomes your column name. We've called our column "Imported Text." Now click **Done** in the Advanced Editor, and then **Close & Apply**.

In Power BI Desktop, we can add the custom visual to our report canvas and drop the "Imported Text" column into the visual, using the Values bucket as shown in Figure 9-8.

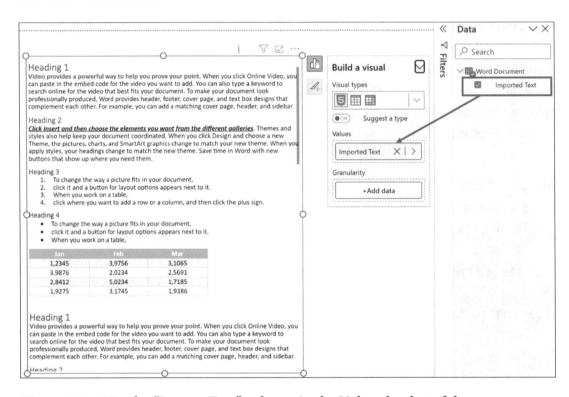

Figure 9-8. *Use the "Import Text" column in the Values bucket of the custom visual*

Note   There are some reports that if you have spaces in your html file name, this will render just a whitespace on refresh. We didn't have this experience as you can see from Figure 9-8, but it may be something to be aware of.

And there we have it, the entire Word document rendered in a Power BI visual on your report canvas.

Q56. How Can I Consolidate Data from Different Excel or CSV Files?

Many of our business processes pre-date Power BI and rely almost exclusively on manipulating Excel data. For this reason, you may be storing data in separate Excel or CSV files that need to be consolidated for analysis purposes. For example, it could be that each year's sales data has been divided into different Excel or CSV files accordingly and each of these files sits in the same folder on your system or in a SharePoint folder. You now need to analyze your sales across all these year. You could, of course, open Excel and do a manual copy-and-paste job to consolidate the data. However, what's going to happen when you have another year's worth of data added to the folder? You will have to repeat this process each time.

Power Query can come to your rescue because we can connect directly to a folder that will then import all the files in that folder from which you can extract the data you need. When you add more files to the folder, you can simply refresh the data in Power BI.

If you want to consolidate data in this way, in Power BI Desktop, you will find both the Folder and SharePoint folder connections on the Get Data list (see Figure 9-9), and you can now browse for your folder.

Figure 9-9. *The Folder and SharePoint folder connectors*

If you're connecting to a SharePoint folder, you'll be prompted to enter the site URL that contains your SharePoint folder. If you're not sure of this URL, in your browser, find the site that holds the folder, and copy the address from the address bar. If you've not connected to this site before, you will be prompted to enter your credentials, and then you can click the **Connect** button.

In the preview that opens, at the bottom right there are three buttons: **Combine & Transform Data**, **Transform Data**, and **Cancel**. If you are connecting to a SharePoint folder, you may want to filter some of the files that are in the site folder. If this is the case, select the **Transform** button. Once you have filtered files as required, click on the **Combine Files** button to the right of the Content column as shown in Figure 9-10.

	Content	↓↓	A^B_C Name	▼	A^B_C Extension	▼
1	Binary		Sales2020.xlsx		.xlsx	
2	Binary		Sales2021.xlsx		.xlsx	
3	Binary		Sales2022.xlsx		.xlsx	

Figure 9-10. *You can filter out files and then click the Combine button*

If you don't need to filter out files in the folder, click on **Combine & Transform Data**. You must now specify which of the files will be used as the sample. By default, this is the first file in the folder. If you are connecting to CSV files, you can click **OK**. If the folder contains Excel files, under "Parameter1[*n*]," click the sheet name that contains the data to be extracted from each file, as shown in Figure 9-11. You can then click **OK** and the files will be appended.

Combine Files

Select the object to be extracted from each file. Lear

Sample File: First file ▼

🔍

Display Options ▼

▲ 🔲 Parameter1 [1]

    Sheet1

Figure 9-11. *You must specify the data to be used as the sample*

The sample data creates a blueprint for how the files will be appended, and therefore, column names in the remaining files must match the column names of the sample file. For Excel files, if you add additional columns to the sample file, these columns will be imported on refresh. However, if you add additional columns to any of the remaining files, these will not be imported. Another important factor is that if connecting to Excel files, table and sheet names must be consistent in all the files.

If you are connecting to CSV files, adding additional columns to the sample file is a little more problematic. This is because, in the "Transform File" function that is generated by Power Query, the M code has a parameter that specifies the number of columns to be imported. If you delete this parameter, this will solve the problem of adding additional columns to the sample file; see Figure 9-12.

```
= (Parameter2) => let
        Source = Csv.Document(Parameter2,[Delimiter=",", Columns=3 Encoding=1252,
            QuoteStyle=QuoteStyle.None]),
        #"Promoted Headers" = Table.PromoteHeaders(Source, [PromoteAllScalars=true])
        in
```

Figure 9-12. *Delete the parameter that limits the columns to be imported*

The only downside to connecting to a folder is that Power Query clutters the Queries pane by generating additional groups and queries.

Q57. Why Would I Need to Unpivot Data?

The reason that this question is frequently posed is that often the requirement to unpivot data is not always apparent. For example, for some reason known only to themselves, Excel users, since time immemorial, have preferred to structure their data as depicted in Figure 9-13.

	A	B	C	D	E	F	G
1		Jan	Feb	Mar	Apr	May	Jun
2	Abel	$2,817	$1,682	$1,900	$1,414	$3,243	$1,781
3	Blanchet	$2,288	$3,199	$4,729	$4,257	$3,960	$1,781
4	Leblanc	$2,286	$1,104	$3,033	$2,040	$2,609	$2,379
5	Reyer	$3,185	$1,094	$2,546	$4,649	$4,456	$2,379
6	Charron	$2,077	$2,494	$2,304	$3,529	$4,883	$3,663
7	Garvey	$2,077	$2,494	$2,304	$3,529	$4,883	$3,020
8	Jordan	$1,466	$3,734	$3,065	$3,032	$4,918	$2,212
9	Denis	$1,466	$3,734	$3,065	$3,032	$4,918	$2,596
10							

Figure 9-13. *Crosstab data in Excel*

I've never fathomed why data arranged in this way is ever a good idea. This layout has a number of descriptions; "crosstab," "matrix," and "pivoted" are some that are applied to it. The trouble is that apart from aggregating the data by the row and column labels, there's little further analysis that you can perform, for example, finding percentages or comparing variances. What we need here is an Excel pivot table, but you can't generate a pivot table from "crosstab" styled data. The same is true when this data arrives in Power BI. We will not be able to work with it as expected to find the insights we're after. For example, consider Figure 9-14 where the Excel data depicted in Figure 9-13 has been imported into Power BI. However, if this data is to be analyzed, it is not organized correctly. You can see that each monthly column label presents itself accordingly.

Figure 9-14. *The crosstab data from Excel does not present itself correctly in Power BI*

We can construct a clustered chart from this data (see Figure 9-15), but it's not a pretty sight; we have far too many values being plotted on the Y-axis.

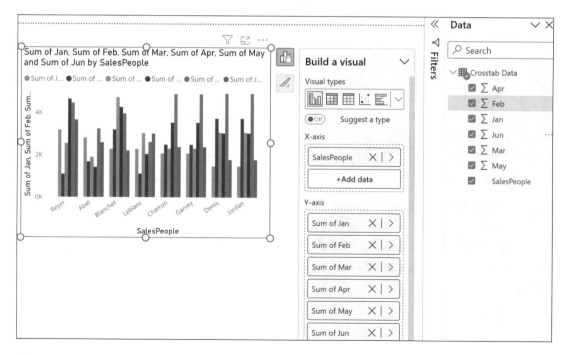

Figure 9-15. *A column chart generated from the crosstab data plots too many values on the Y-axis*

And what if we wanted to generate a card visual to show the grand total? How would we do this? We would have to create a DAX measure that added together the values in each monthly column! This is not a way forward.

If this data is to be sensibly analyzed, it must be re-structured whereby there are repeating rows for each combination of salesperson, month, and value. Then we can show the monthly value for each of these combinations in a column of its own. We would refer to such a structure as "tabular." All data used in Power BI must have a tabular structure. We can use Power Query to reshape the data accordingly, and this is known as "unpivoting data."

To do this, in Power Query, we must select the SalesPerson column (that has been renamed from "Column1"), and then select **Unpivot Other Columns** from the **Transform** tab, Figure 9-16.

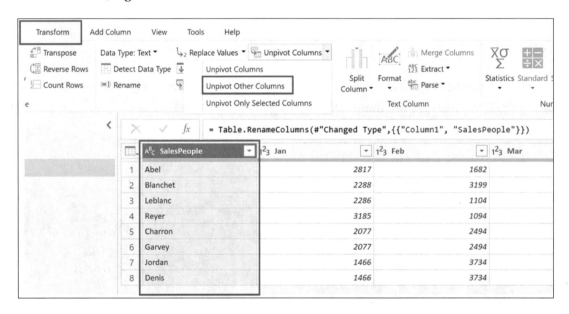

Figure 9-16. *Unpivoting columns*

We can then rename the "Attribute" column, "Month," and the "Value" column, "Sales," for example. You can see in Figure 9-17 how the data has been reshaped into a tabular structure.

	ABC SalesPeople	ABC Month	123 Sales
1	Abel	Jan	2817
2	Abel	Feb	1682
3	Abel	Mar	1900
4	Abel	Apr	1414
5	Abel	May	3243
6	Abel	Jun	1781
7	Blanchet	Jan	2288
8	Blanchet	Feb	3199
9	Blanchet	Mar	4729
10	Blanchet	Apr	4257
11	Blanchet	May	3960
12	Blanchet	Jun	1781
13	Leblanc	Jan	2286
14	Leblanc	Feb	1104
15	Leblanc	Mar	3033
16	Leblanc	Apr	2040
17	Leblanc	May	2609

Figure 9-17. *The data now has a tabular structure*

To sort the month names correctly, you must add a month number column. You can do this using a Conditional Column or a Column From Examples, buttons for which are on the **Add Column** tab. To sort the month names by the month number, refer to QU24 "Why Do I Need a Date Table". Once this data is loaded, we can plot the data more sensibly in any visual we choose.

The problem with the example we've just explored, however, is that often the original layout in Excel is more complex than a simple crosstab. Consider Figure 9-18 where above the month labels, we have quarter labels and above these, labels for each year. There are also now two row labels, "Salesperson" and "Group."

	A	B	C	D	E	F	G	H	I	J	K	L	M	N
1						2021						2022		
2				Qtr1			Qtr2			Qtr1			Qtr2	
3	Salesperson	Group	Jan	Feb	Mar	Apr	May	Jun	Jan	Feb	Mar	Apr	May	Jun
4	Abel	A	$2,136	$3,070	$2,535	$1,579	$1,140	$1,037	$1,244	$1,873	$2,328	$3,847	$2,042	$2,504
5	Blanchet	A	$1,300	$3,490	$1,578	$2,927	$1,088	$1,444	$2,762	$3,715	$2,705	$2,087	$2,141	$1,296
6	Leblanc	A	$2,567	$2,835	$2,400	$1,311	$2,338	$2,160	$2,840	$2,600	$3,583	$3,516	$3,215	$2,243
7	Reyer	A	$3,164	$3,904	$1,305	$1,052	$3,117	$3,811	$2,384	$2,993	$3,946	$3,950	$3,994	$1,968
8	Charron	A	$1,189	$3,308	$3,385	$2,994	$1,985	$1,140	$2,387	$2,071	$2,055	$3,233	$1,379	$2,091
9	Garvey	A	$1,324	$2,421	$3,784	$2,073	$1,135	$3,278	$1,344	$1,464	$3,825	$1,975	$3,697	$3,632
10	Jordan	A	$2,547	$1,859	$1,625	$2,342	$2,249	$2,789	$1,920	$1,102	$2,989	$1,641	$3,954	$2,844
11	Denis	A	$1,502	$2,228	$1,515	$3,241	$1,186	$1,575	$3,936	$2,731	$3,233	$1,312	$2,963	$1,804
12	Abel	B	$2,883	$1,156	$2,828	$2,969	$2,660	$2,924	$2,997	$1,346	$2,376	$1,659	$1,722	$2,448
13	Blanchet	B	$2,665	$1,023	$2,212	$1,700	$1,555	$2,251	$1,581	$1,130	$1,458	$1,653	$2,257	$2,767
14	Leblanc	B	$1,411	$2,463	$1,871	$1,735	$2,831	$1,012	$1,169	$1,418	$1,708	$1,914	$2,297	$1,654
15	Reyer	B	$1,138	$2,031	$2,033	$1,874	$2,906	$1,464	$1,791	$1,910	$1,184	$2,678	$1,949	$2,251
16	Charron	B	$1,980	$1,784	$2,124	$1,688	$2,405	$2,852	$1,245	$1,985	$1,859	$1,225	$2,363	$1,792
17	Garvey	B	$1,343	$2,019	$2,965	$2,325	$2,969	$1,967	$2,243	$1,165	$2,821	$2,185	$1,206	$2,620
18	Jordan	B	$1,890	$2,646	$2,921	$2,181	$2,348	$1,319	$2,115	$2,613	$2,721	$1,922	$2,391	$1,560
19	Denis	B	$2,723	$1,784	$1,167	$2,178	$1,749	$1,716	$1,603	$1,424	$2,769	$2,474	$2,013	$2,444

Figure 9-18. *Crosstab data in Excel can be complex*

However, we can bring Power Query to the rescue again to re-model this data into a tabular layout.

When the data is imported into Power Query, it will present itself as depicted in Figure 9-19 which is much the same as its layout in Excel.

	Column1	Column2	Column3	Column4	Column5
1	null	null	2021	null	
2	null	null	Qtr1	null	
3	Salesperson	Group	Jan	Feb	Mar
4	Abel	A	2136	3070	
5	Blanchet	A	1300	3490	
6	Leblanc	A	2567	2835	
7	Reyer	A	3164	3904	
8	Charron	A	1189	3308	
9	Garvey	A	1324	2421	
10	Jordan	A	2547	1859	
11	Denis	A	1502	2228	
12	Abel	B	2883	1156	
13	Blanchet	B	2665	1023	
14	Leblanc	B	1411	2463	
15	Reyer	B	1138	2031	
16	Charron	B	1980	1784	
17	Garvey	B	1343	2019	
18	Jordan	B	1890	2646	
19	Denis	B	2723	1784	

Figure 9-19. *How the Excel data will first present itself in Power Query*

The starting point for transforming this data is to transpose the data using the button on the **Transform** tab; this will swap the rows and columns around; see Figure 9-20.

	Column1	Column2	Column3	Column4	Column5
1	null	null	Salesperson	Abel	Blanchet
2	null	null	Group	A	A
3	2021	Qtr1	Jan	2136	
4	null	null	Feb	3070	
5	null	null	Mar	2535	
6	null	Qtr2	Apr	1579	
7	null	null	May	1140	
8	null	null	Jun	1037	
9	2022	Qtr1	Jan	1244	
10	null	null	Feb	1873	
11	null	null	Mar	2328	
12	null	Qtr2	Apr	3847	
13	null	null	May	2042	
14	null	null	Jun	2504	

Figure 9-20. *Transposing the data*

We can now make use of the "Fill Down" transformation to populate the missing year and quarter labels. Select "Column1" and "Column2," and using the dropdown on the **Fill** button on the **Transform** tab, we can **Fill Down**; see Figure 9-21.

	ABC 123 Column1		ABC 123 Column2	
1	null		null	
2	null		null	
3	2021	Qtr1		
4	2021	Qtr1		
5	2021	Qtr1		
6	2021	Qtr2		
7	2021	Qtr2		
8	2021	Qtr2		
9	2022	Qtr1		
10	2022	Qtr1		
11	2022	Qtr1		
12	2022	Qtr2		
13	2022	Qtr2		
14	2022	Qtr2		

Figure 9-21. *Filling down to fill in the missing year and quarter values*

The next step is to generate the values that will eventually become the column headers. To do this, we must select "Column1," "Column2," and "Column3," and using the **Merge Columns** button on the **Transform** tab, merge this data using a space as a separator. The result of this transformation can be seen in Figure 9-22.

	ABC Merged		ABC 123 Column4
1	Salesperson		Abel
2	Group		A
3	2021 Qtr1 Jan		
4	2021 Qtr1 Feb		
5	2021 Qtr1 Mar		
6	2021 Qtr2 Apr		
7	2021 Qtr2 May		
8	2021 Qtr2 Jun		
9	2022 Qtr1 Jan		
10	2022 Qtr1 Feb		
11	2022 Qtr1 Mar		
12	2022 Qtr2 Apr		
13	2022 Qtr2 May		
14	2022 Qtr2 Jun		

Figure 9-22. *Merging the columns*

Now we are able to convert these row values into column headings by transposing the data again and using the "**Use First Row as Headers**" button on the **Transform** tab. The data now presents itself as shown in Figure 9-23.

	ABC 123 Salesperson	ABC 123 Group	ABC 123 2021 Qtr1 Jan	ABC 123 2021 Qtr1 Feb	ABC 123
1	Abel	A	2136	3070	
2	Blanchet	A	1300	3490	
3	Leblanc	A	2567	2835	
4	Reyer	A	3164	3904	
5	Charron	A	1189	3308	
6	Garvey	A	1324	2421	
7	Jordan	A	2547	1859	
8	Denis	A	1502	2228	
9	Abel	B	2883	1156	
10	Blanchet	B	2665	1023	
11	Leblanc	B	1411	2463	
12	Reyer	B	1138	2031	
13	Charron	B	1980	1784	

Figure 9-23. *Transposing the data and promoting the first row to headers*

At last, we're ready to unpivot the data. The "Salesperson" and "Group" columns are correctly structured. Therefore, we can select these, and using "**Unpivot Other Columns**" we can unpivot the columns containing the year, quarter, and month labels. This step is shown in Figure 9-24.

	ABC Salesperson	ABC Group	ABC Attribute	123 Value
1	Abel	A	2021 Qtr1 Jan	2136
2	Abel	A	2021 Qtr1 Feb	3070
3	Abel	A	2021 Qtr1 Mar	2535
4	Abel	A	2021 Qtr2 Apr	1579
5	Abel	A	2021 Qtr2 May	1140
6	Abel	A	2021 Qtr2 Jun	1037
7	Abel	A	2022 Qtr1 Jan	1244
8	Abel	A	2022 Qtr1 Feb	1873
9	Abel	A	2022 Qtr1 Mar	2328
10	Abel	A	2022 Qtr2 Apr	3847
11	Abel	A	2022 Qtr2 May	2042
12	Abel	A	2022 Qtr2 Jun	2504
13	Blanchet	A	2021 Qtr1 Jan	1300
14	Blanchet	A	2021 Qtr1 Feb	3490
15	Blanchet	A	2021 Qtr1 Mar	1578
16	Blanchet	A	2021 Qtr2 Apr	2927
17	Blanchet	A	2021 Qtr2 May	1088
18	Blanchet	A	2021 Qtr2 Jun	1444

Figure 9-24. *Unpivoting the columns containing year, qtr, and month labels*

Now all that remains for us to do is to split the "Attribute" column using the **Split Column** button on the **Transform** tab, splitting by the space delimiter. After renaming the columns, we can click the **Close & Apply** button. Figure 9-25 shows the final unpivoted data in Data view.

Salesperson ▼	Group ▼	Year ▼	Qtr ▼	Month ▼	Sales ▼
Abel	A	2021	Qtr1	Jan	2136
Abel	A	2021	Qtr1	Feb	3070
Abel	A	2021	Qtr1	Mar	2535
Abel	A	2021	Qtr2	Apr	1579
Abel	A	2021	Qtr2	May	1140
Abel	A	2021	Qtr2	Jun	1037
Abel	A	2022	Qtr1	Jan	1244
Abel	A	2022	Qtr1	Feb	1873
Abel	A	2022	Qtr1	Mar	2328
Abel	A	2022	Qtr2	Apr	3847
Abel	A	2022	Qtr2	May	2042
Abel	A	2022	Qtr2	Jun	2504
Blanchet	A	2021	Qtr1	Jan	1300
Blanchet	A	2021	Qtr1	Feb	3490
Blanchet	A	2021	Qtr1	Mar	1578

Figure 9-25. *The restructured data*

Using much the same steps as outlined here, you can take complex crosstab-style data in Excel and convert it into tabular data; see Figure 9-26.

	A	B	C	D	E	F	G	H	I	J	K	L	M	N	O
1								Acme Ltd							
2						USA						London			
3						2021						2022			
4					Qtr1			Qtr2			Qtr1			Qtr2	
5	Salesperson	Product	Group	Jan	Feb	Mar	Apr	May	Jun	Jan	Feb	Mar	Apr	May	Jun
6	Abel	Merlot	A	$2,136	$3,070	$2,535	$1,579	$1,140	$1,037	$1,244	$1,873	$2,328	$3,847	$2,042	$2,504
7	Blanchet	Sauvignon Bl	A	$1,300	$3,490	$1,578	$2,927	$1,088	$1,444	$2,762	$3,715	$2,705	$2,087	$2,141	$1,296
8	Leblanc	Chianti	A	$2,567	$2,835	$2,400	$1,311	$2,338	$2,160	$2,840	$2,600	$3,583	$3,516	$3,215	$2,243
9	Reyer	Malbec	A	$3,164	$3,904	$1,305	$1,052	$3,117	$3,811	$2,384	$2,993	$3,946	$3,950	$3,994	$1,968
10	Charron	Riocha	A	$1,189	$3,308	$3,385	$2,994	$1,985	$1,140	$2,387	$2,071	$2,055	$3,233	$1,379	$2,091
11	Garvey	Champagne	A	$1,324	$2,421	$3,784	$2,073	$1,135	$3,278	$1,344	$1,464	$3,825	$1,975	$3,697	$3,632
12	Jordan	Bordeaux	A	$2,547	$1,859	$1,625	$2,342	$2,249	$2,789	$1,920	$1,102	$2,989	$1,641	$3,954	$2,844
13	Denis	Pinot Grigio	A	$1,502	$2,228	$1,515	$3,241	$1,186	$1,575	$3,936	$2,731	$3,233	$1,312	$2,963	$1,804
14	Abel	Merlot	B	$2,883	$1,156	$2,828	$2,969	$2,660	$2,924	$2,997	$1,346	$2,376	$1,659	$1,722	$2,448
15	Blanchet	Sauvignon Bl	B	$2,665	$1,023	$2,212	$1,700	$1,555	$2,251	$1,581	$1,130	$1,458	$1,653	$2,257	$2,767
16	Leblanc	Chianti	B	$1,411	$2,463	$1,871	$1,735	$2,831	$1,012	$1,169	$1,418	$1,708	$1,914	$2,297	$1,654
17	Reyer	Malbec	B	$1,138	$2,031	$2,033	$1,874	$2,906	$1,464	$1,791	$1,910	$1,184	$2,678	$1,949	$2,251
18	Charron	Riocha	B	$1,980	$1,784	$2,124	$1,688	$2,405	$2,852	$1,245	$1,985	$1,859	$1,225	$2,363	$1,792

Salesperson	Product	Group	Customer	Country	Year	Qtr	Month	Sales
Leblanc	Chianti	B	Acme Ltd	USA	2021	Qtr2	Jun	1012
Blanchet	Sauvignon Blanc	B	Acme Ltd	USA	2021	Qtr1	Feb	1023
Abel	Merlot	A	Acme Ltd	USA	2021	Qtr2	Jun	1037
Reyer	Malbec	A	Acme Ltd	USA	2021	Qtr2	Apr	1052
Blanchet	Sauvignon Blanc	A	Acme Ltd	USA	2021	Qtr2	May	1088
Jordan	Bordeaux	A	Acme Ltd	London	2022	Qtr1	Feb	1102
Blanchet	Sauvignon Blanc	B	Acme Ltd	London	2022	Qtr1	Feb	1130
Garvey	Champagne	A	Acme Ltd	USA	2021	Qtr2	May	1135
Reyer	Malbec	B	Acme Ltd	USA	2021	Qtr1	Jan	1138
Abel	Merlot	A	Acme Ltd	USA	2021	Qtr2	May	1140
Charron	Riocha	A	Acme Ltd	USA	2021	Qtr2	Jun	1140
Abel	Merlot	B	Acme Ltd	USA	2021	Qtr1	Feb	1156
Garvey	Champagne	B	Acme Ltd	London	2022	Qtr1	Feb	1165
Denis	Pinot Grigio	B	Acme Ltd	USA	2021	Qtr1	Mar	1167

Figure 9-26. *Restructuring data from Excel using the Unpivot transformation*

The requirement to unpivot data mostly stems from legacy structures of Excel data, in that Excel users appear to prefer crosstab style layouts for their data. You will also often find similar layouts if you connect to data from websites. In Power BI, you can only successfully analyze data that has a tabular structure, but you can use Power Query to reshape the data accordingly.

Q58. Why Would I Need to Pivot Data?

Let's now address the opposite transformation from unpivoting data and that is the requirement to pivot it. Just as with unpivoting data, the need to pivot it stems mostly from legacy Excel layouts. Consider Figure 9-27 where the data is neither laid out with a crosstab structure nor with a tabular structure.

	A	B	C	D
1	NAME	TYPE	YEAR	VALUE
2	Abel	Actual	2016	$1,938
3		Target	2016	$1,703
4		Actual	2017	$1,703
5		Target	2017	$1,232
6	Blanchet	Actual	2016	$1,761
7		Target	2016	$1,640
8		Actual	2017	$1,239
9		Target	2017	$1,888
10	Leblanc	Actual	2016	$1,193
11		Target	2016	$1,911
12		Actual	2017	$1,272
13		Target	2017	$1,590

Figure 9-27. Data in Excel that is neither crosstab nor tabular

We can see that the "TYPE" data does not sit comfortably as a column within the table. The obvious problem is that the values are repeated, but it also flags up the issue where there are two different types of data in the "VALUE" column, that is, "Actual" values and "Target" values. To analyze this data, the layout we require is shown in Figure 9-28 where the "Actual" and "Target" values have been brought into columns of their own.

NAME ▾	YEAR ▾	ACTUAL ▾	TARGET ▾
Abel	2016	$1,938	$1,703
Abel	2017	$1,703	$1,232
Blanchet	2016	$1,761	$1,640
Blanchet	2017	$1,239	$1,888
Charron	2016	$1,411	$1,492
Charron	2017	$1,220	$1,945
Denis	2016	$1,184	$1,190
Denis	2017	$1,496	$1,598
Garvey	2016	$1,338	$1,689
Garvey	2017	$1,017	$1,591
Jordan	2016	$1,721	$1,965

Figure 9-28. The Excel data is now tabular

To restructure the data accordingly, we must use the pivot transformation that moves labels that sit in rows into headers so that they correctly label their corresponding values. In Power Query, after filling down on the NAME column to populate the null values (using the **Fill** button dropdown on the **Transform** tab), we must select the TYPE column and on the **Transform** tab and use the **Pivot Column** button. In the Pivot Column window, we would then select the "VALUE" column in the **Values Column** dropdown,

Figure 9-29. *In the Pivot Column window, select the column that holds the values to be pivoted*

In the example we've just explored, we pivoted a numeric column under the "Actual" and "Target" labels. However, there is often a requirement to pivot text values instead. If we take a look at the data in Figure 9-30, we can see that we must pivot the INFO column to bring the values in the DETAILS column under the corresponding labels. However, the values in the DETAILS column are both text and numeric.

D	E	F
WINE	**INFO**	**DETAILS**
Bordeaux	TYPE	Red
Bordeaux	WINE COUNTRY	France
Bordeaux	PRICE PER CASE	$75.00
Champagne	TYPE	White
Champagne	WINE COUNTRY	Italy
Champagne	PRICE PER CASE	$100.00
Chardonnay	TYPE	White
Chardonnay	WINE COUNTRY	Germany
Chardonnay	PRICE PER CASE	$75.00
Malbec	TYPE	Red
Malbec	WINE COUNTRY	Germany
Malbec	PRICE PER CASE	$85.00
Grenache	TYPE	Red
Grenache	WINE COUNTRY	France

Figure 9-30. *The DETAILS column holds text and numeric values to be pivoted under the labels in the INFO column*

To do this, we must change the Advanced options in the Pivot Column window as shown in Figure 9-31, making sure we select "Don't Aggregate."

Pivot Column

Use the names in column "INFO" to create new columns.

Values Column ⓘ

DETAILS	▾

▲ Advanced options

Aggregate Value Function

Don't Aggregate	▾

Learn more about Pivot Column

Figure 9-31. *To pivot text data, use the "Don't Aggregate" option*

Once we've done this, the data will now present itself as tabular data with the correct columns, as shown in Figure 9-32.

WINE	▾	TYPE	▾	WINE COUNTRY	▾	PRICE PER CASE	▾
Bordeaux		Red		France		$75.00	
Champagne		White		Italy		$100.00	
Chardonnay		White		Germany		$75.00	
Chenin Blanc		White		France		$50.00	
Chianti		Red		Italy		$40.00	
Grenache		Red		France		$30.00	
Lambrusco		White		Italy		$20.00	
Malbec		Red		Germany		$85.00	
Merlot		Red		France		$39.00	
Piesporter		White		France		$30.00	
Pinot Grigio		White		Italy		$30.00	
Rioja		Red		Italy		$45.00	

Figure 9-32. *The Excel data is now tabular*

However, incorrectly structured data sometimes presents itself where the solution is to perform a pivot transformation just won't work. If we look at the Excel data in Figure 9-33, for example, it would seem that the solution to convert the data into a tabular layout would be to pivot the INFO column.

	A	B
1	**INFO**	**DETAILS**
2	WINE	Bordeaux
3	TYPE	Red
4	WINE COUNTRY	France
5	PRIC EPER CASE	$75.00
6	WINE	Champagne
7	TYPE	White
8	WINE COUNTRY	Italy
9	PRIC EPER CASE	$100.00
10	WINE	Chardonnay
11	TYPE	White

Figure 9-33. *This Excel data must be pivoted to make it tabular*

If we attempt to do so, using the DETAILS column as the values column, set to "Don't Aggregate" as in Figure 9-31, we get errors; see Figure 9-34.

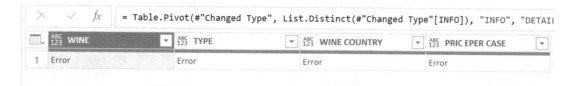

Figure 9-34. *Attempting to pivot the INFO column results in errors*

The problem is this: when you pivot data, you must specify the column to be pivoted and the column that holds the values to be populated into the pivoted column. The other columns in the table will be used to group the data accordingly. In Figure 9-28, for example, the data has been grouped by NAME and YEAR. With the data in Figure 9-33, where we're pivoting the INFO column and the values to be pivoted are in the DETIALS column, there are no additional columns by which to group the data. For this reason, we must generate values by which to group each set of four rows in the INFO column, that is, WINE, TYPE, WINE COUNTRY, and PRICE PER CASE.

Before pivoting the data, therefore, we must add an **Index Column**, by using the button on the **Add Column** tab. Ensure that you start the numbering at "**0**." We can now divide each value in the index column by 4 because we want to group four rows at a time. We can then use the integer part of these divisions, ignoring the remainder. Power Query gives us a transformation for doing this. With the Index column selected, use the dropdown on the **Standard** button on the **Transform** tab, and select **Integer-Divide**, entering 4 as the value by which to divide. The transformed Index column is shown in Figure 9-35.

	A^B_C INFO	ABC 123 DETAILS	1²3 Index
1	WINE	Bordeaux	0
2	TYPE	Red	0
3	WINE COUNTRY	France	0
4	PRIC EPER CASE	75	0
5	WINE	Champagne	1
6	TYPE	White	1
7	WINE COUNTRY	Italy	1
8	PRIC EPER CASE	100	1
9	WINE	Chardonnay	2
10	TYPE	White	2
11	WINE COUNTRY	Germany	2
12	PRIC EPER CASE	75	2
13	WINE	Malbec	3
14	TYPE	Red	3

Figure 9-35. *Using the Integer-Division column to group the rows*

This column can now be used in the pivot transformation. If we select the INFO column to be pivoted and specify the DETAILS column as the value, setting this to "Don't Aggregate," this will now work.

In the last example, where we're answering the question, "Why Would I Need to Pivot Data?" the data is simply a column of values; see Figure 9-36.

	A	B
1	**DETAILS**	
2	Bordeaux	
3	Red	
4	France	
5	£75.00	
6	Champagne	
7	White	
8	Italy	
9	£100.00	
10	Chardonnay	
11	White	
12	Germany	
13	£75.00	
14	Malbec	
15	Red	

Figure 9-36. *A simple column of data in Excel*

To analyze this data in Power BI, we must transform it into a table comprising columns and rows, as illustrated in Figure 9-37.

WINE	TYPE	WINE COUNTRY	PRICE PER CASE
Bordeaux	Red	France	$75.00
Champagne	White	Italy	$100.00
Chardonnay	White	Germany	$75.00
Chenin Blanc	White	France	$50.00
Chianti	Red	Italy	$40.00
Grenache	Red	France	$30.00
Lambrusco	White	Italy	$20.00
Malbec	Red	Germany	$85.00
Merlot	Red	France	$39.00
Piesporter	White	France	$30.00
Pinot Grigio	White	Italy	$30.00

Figure 9-37. *The data in Figure 9-36, transformed into a table*

The issue with the data in Figure 9-36 is that there is no column by which to pivot the data (i.e., no column that provides the column headers), nor is there any column by which to group the data. Therefore, we must generate both these columns ourselves. We learned in the previous example that to group the data we can use an integer division by 4 on an Index column. After creating this Index column, we must now generate another column by which to pivot the data. We can do this by assigning a number to each row that will then become the new column headers. For example, each product name will be numbered "1," each type (red or white) will be "2," the wine country will be "3," and so on. For this, add another Index column starting at "**0**." With this Index column selected, on the **Transform** tab using the **Standard button** dropdown, select **Modulo**, using "4" as the modulo value. We now have columns by which to group and pivot the data in the DETAILS column, as shown in Figure 9-38.

	ABC 123 DETAILS	1²3 Index	1.2 Index.1
1	Bordeaux	0	0
2	Red	0	1
3	France	0	2
4	75	0	3
5	Champagne	1	0
6	White	1	1
7	Italy	1	2
8	100	1	3
9	Chardonnay	2	0
10	White	2	1
11	Germany	2	2
12	75	2	3
13	Malbec	3	0
14	Red	3	1

Figure 9-38. *Generating values by which to group and pivot the data in the DETAILS column*

The "Index.1" column can now be pivoted using the DETAILS column as the Values column and setting the option to "Don't Aggregate"; see Figure 9-39.

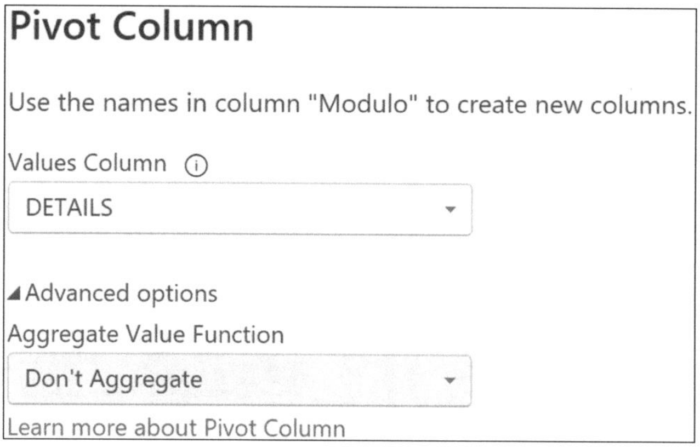

Pivot Column

Use the names in column "Modulo" to create new columns.

Values Column ⓘ

| DETAILS | ▾ |

▴Advanced options

Aggregate Value Function

| Don't Aggregate | ▾ |

Learn more about Pivot Column

Figure 9-39. *Pivoting the "Modulo" column*

The Index column can now be removed, and all that remains to be done is to rename the columns accordingly as they will have been named by the numbers generated by the modulo values. Once you have loaded the query into Power BI, it will be reshaped as depicted in Figure 9-37.

Q59. How Do I Generate a Cartesian Product?

It's sometimes the case that you want to find all possible combinations of entities, for example, all combinations of your products and sales locations. A Cartesian product is a method of joining tables that results in all such possible combinations of pairs of rows. The Cartesian product of two sets A and B is denoted by A × B. It's possible to use the DAX function GENERATE to do this, but if Power Query can do the job, this would always be the preferred tool.

Note   If you want more information as to why Power Query is the preferred tool for table transformations, visit `www.burningsuit.co.uk/blog/where-do-i-make-data-transformations-power-bi-roches-maxim`.

Let's take a very simple example of where a Cartesian product might be useful. We may have a separate list of years and a separate list of months. We would like to generate a table combining all months in all years as depicted in Figure 9-40.

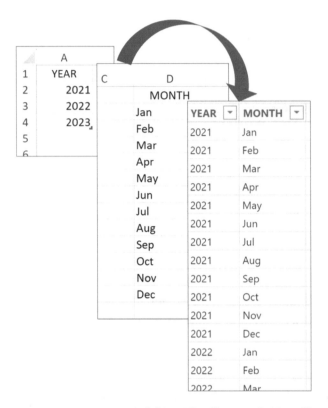

Figure 9-40. *We want to generate a table with all months in all years*

To do this, you must import the list of years and list of months into Power Query as two separate tables.

Note   If you are using the sample data in the "Power Query Data" Excel file, you may need to use the "**Use First Row as Headers**" transformation on the **Home** tab if the column headers sit in the first row.

Ensure you rename the tables "Year" and "Month" accordingly. Then, in Power Query, with the Year table selected, use the **Custom Column** button on the **Add Column** tab. In the formula, reference the Month table, as shown in Figure 9-41.

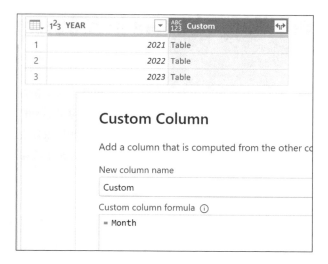

Figure 9-41. *Use a Custom Column and reference the table you want combined in a Cartesian product*

In the Custom Column header, click the double arrow button to expand the table to show the month names, and click **OK**, as shown in Figure 9-42.

Figure 9-42. *Expand the table in the Custom Column*

Once we've generated our table of year and month labels, this table could then be used in further steps to merge with other data. Therefore, to generate the Cartesian product of any number of tables, simply add a Custom Column referencing the table to be merged, and then expand the Custom Column.

Q60. How Can I Merge Values in a Column into a Row?

Consider the Excel data in Figure 9-43 and how it has been reshaped by Power Query. You can see that the children's names in the CHILDREN column have been transformed into a single row of concatenated names against the FIRSTNAME and SURNAME values and the redundant blank cells in these columns have been removed.

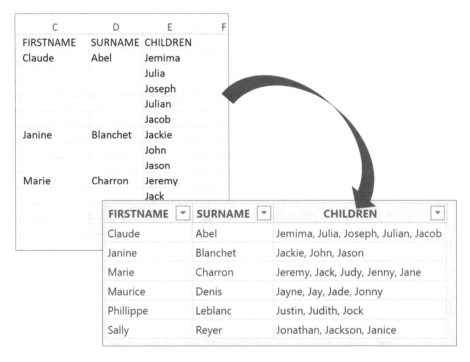

Figure 9-43. *Excel data that has been reshaped by Power Query*

So how do you merge values in a column into a row? Let's start by importing the Excel data into Power Query.

Note   If you are using the sample data in the "Power Query Data" Excel file, you may need to use the "**Use First Row as Headers**" transformation on the **Home** tab if the column headers sit in the first row.

The next step is to select the FIRSTNAME and SURNAME columns, and on the **Transform** tab, use the **Fill Down** option on the dropdown of the **Fill** button. With the FIRSTNAME and SURNAME columns still selected, click the **Group By** button on the **Transform** tab, and complete the entries as shown in Figure 9-44.

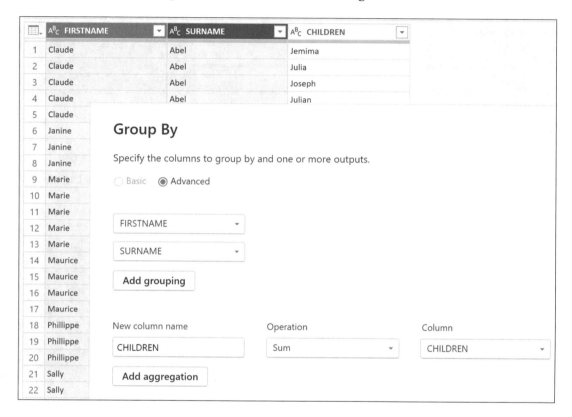

Figure 9-44. *Grouping by FIRSTNAME and SURNAME*

We will name the new column "CHILDREN." In the "Operation" dropdown, you can select any aggregate function. This is because the CHILDREN column has been selected in the "Column" dropdown, and because you can't aggregate text values, we will get an error when we click the OK button; see Figure 9-45.

	A^Bc FIRSTNAME		A^Bc SURNAME		A^Bc CHILDREN	
1	Claude		Abel		Error	
2	Janine		Blanchet		Error	
3	Marie		Charron		Error	
4	Maurice		Denis		Error	
5	Phillippe		Leblanc		Error	
6	Sally		Reyer		Error	

Figure 9-45. *An error is returned in the CHILDREN column*

To resolve the error, make sure you've selected the "Grouped Rows" step in the **APPLIED STEPS** pane. Now, on the **View** tab, turn on the **Formula Bar** to see the M code for the "Grouped Rows" step; see Figure 9-46.

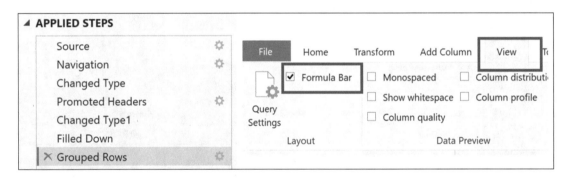

Figure 9-46. *Turn on the Formula Bar to see the M code for the Grouped Rows step*

Now in the Formula bar, edit the "M" code as shown in Figure 9-47. This statement

```
List.Sum([CHILDREN])
```

is edited to

```
Text.Combine([CHILDREN], ", ")
```

Note the last parameter of "Text.Combine" determines the separator for the concatenated values.

```
= Table.Group(#"Filled Down", {"FIRSTNAME", "SURNAME"}, {{"CHILDREN",
    each List.Sum([CHILDREN]), type nullable text}})

= Table.Group(#"Filled Down", {"FIRSTNAME", "SURNAME"}, {{"CHILDREN",
    each Text.Combine([CHILDREN],", "), type nullable text}})
```

Figure 9-47. *Edit the M code in the formula bar*

It might seem strange there is not a generic transformation in Power Query for concatenating values in this way. In DAX there is a function called CONCATENATEX that will do this job. However, editing M code as explained here is perhaps easier than getting to grips with DAX.

Q61. Can I Use Values That Don't Match to Merge Data?

Consider Figure 9-48 where we have our products listed in the PRODUCT column of the Products and Sales tables. We must calculate revenue value by multiplying the price in the Products table by the quantity in the Sales table. To do this we must merge the rows from the Products table with the rows of the Sales table. In the PRODUCT column of the Sales table, the values have been misspelled, so how will we do this when the values in the PRODUCT columns don't match?

Products

	A^BC PRODUCT	1²3 PRICE	
1	Bordeaux	75	
2	Champagne	200	
3	Chardonnay	150	
4	Malbec	85	
5	Grenache	30	
6	Piesporter	125	
7	Chianti	40	
8	Pinot Grigio	30	
9	Merlot	39	
10	Sauvignon Blanc	40	
11	Rioja	45	
12	Chenin Blanc	50	
13	Shiraz	78	
14	Lambrusco	20	

Sales

NO	A^BC PRODUCT	1²3 QUANTITY	
1	Chiati	145	
2	s blanc	111	
3	Gren ache	99	
4	Bordux	110	
5	grenache	291	
6	P Grigo	233	
7	Pino Grigio	187	
8	Chiantie	267	
9	Chardonay	99	
02/01/2018	10	hiraz	132
02/01/2018	11	Psporter	108
03/01/2018	12	Champane	273
03/01/2018	13	CheEninBlanc	90
04/01/2018	14	sauvig blanc	154
04/01/2018	15	Suavignon Blanc	136
04/01/2018	16	Pinot Grijio	268
04/01/2018	17	Cpagne	98
05/01/2018	18	Shiraz	186

Figure 9-48. *The values in the PRODUCT columns don't match*

We can use Power Query's Fuzzy Merge option to do this and this is how.

Note   If you are using the sample data in the "Power Query Data" Excel file, rename the tables, "Sales" and "Products."

Making sure the Sales table is selected, on the **Home** tab, we can select the **Merge Queries** option. In the **Merge** window, select the Products table as the table to merge with the Sales table. Click the PRODUCT column in both tables to identify the values to be matched. Note how we are told how many matching values have been found; see Figure 9-49.

Figure 9-49. *Merging the Products table with the Sales table using the PRODUCT columns – note the number of matching values*

Now turn on the "**Use fuzzy matching to perform the merge**" check box. Note how the number of matches increases. In our sample data, this was from **3,008** to **3,019**. We can now click **OK** in the Merge window, and a new column will be generated labeled, "Product.1." Click the expand button to the right of the column header, and click **OK**; see Figure 9-50.

Figure 9-50. *Expand the merged data*

However, there are still a number of non-matching values that have been populated with nulls; see Figure 9-51.

NO	A^B_C PRODUCT	1²3 QUANTITY	A^B_C Product.1.PRODUCT	1²
1	Chiati	145	Chianti	
2	s blanc	111		null
3	Gren ache	99	Grenache	
4	Bordux	110	Bordeaux	
5	grenache	291	Grenache	
6	P Grigo	233		null
7	Pino Grigio	187	Pinot Grigio	
8	Chiantie	267	Chianti	
9	Chardonay	99	Chardonnay	
10	hiraz	132	Shiraz	
11	Psporter	108	Piesporter	
12	Champane	273	Champagne	
13	CheEninBlanc	90		null
14	sauvig blanc	154		null
15	Suavignon Blanc	136	Sauvignon Blanc	
16	Pinot Grijio	268	Pinot Grigio	
17	Cpagne	98		null

Figure 9-51. *The Fuzzy matching has not matched all the values in the PRODUCT column*

What we can do here is decrease the similarity threshold of the fuzzy merge. In the **APPLIED STEPS** pane, edit the "Merged Queries" step by clicking on the gear wheel to the right. Expand the "**Fuzzy matching options**," and in the "**Similarity threshold**" option, enter **0** which will cause any values with any level of similarity to match. After clicking **OK**, you can then select the "Expanded Products" step to see the new merged results. What you will notice now is that there are duplicate rows where alternative matched results have been generated. For example, "s blanc" has been matched to both "Sauvignon Blanc" and "Chenin Blanc." Again we can edit the "Merged Queries" step and Fuzzy matching options. In the "**Maximum number of matches**" option, enter **1**, to ensure that only one row is returned for each match. This improves the number of matches but we're not quite there yet. There are two values that have not been matched: "porter" and "Cpagne." What we must do here is generate a transformation table to map these values to the correct product names.

To do this, click the **Enter Data** button on the **Home** tab, and create a table of values as shown in Figure 9-52. It's mandatory that the column names are "From" and "To." The "From" values are not case-sensitive. We called our table "Transformation" but it can have any name.

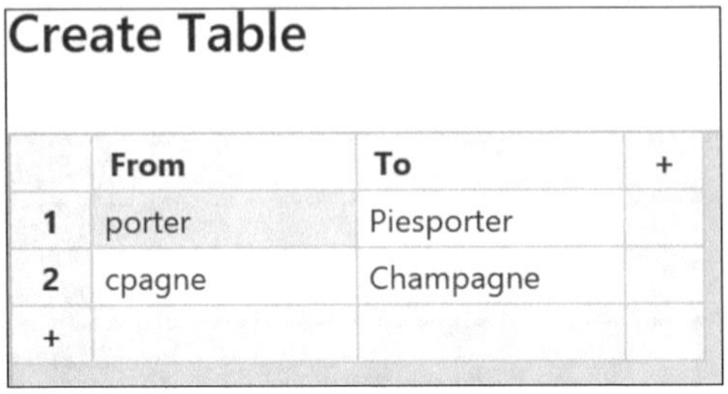

Create Table			
	From	**To**	+
1	porter	Piesporter	
2	cpagne	Champagne	
+			

Figure 9-52. *The transformation table for the fuzzing merge*

Once you have created the transformation table, you can edit the "Merged Queries" step and the Fuzzing matching options to include the transformation table. The Fuzzing matching options that we have edited are shown in Figure 9-53.

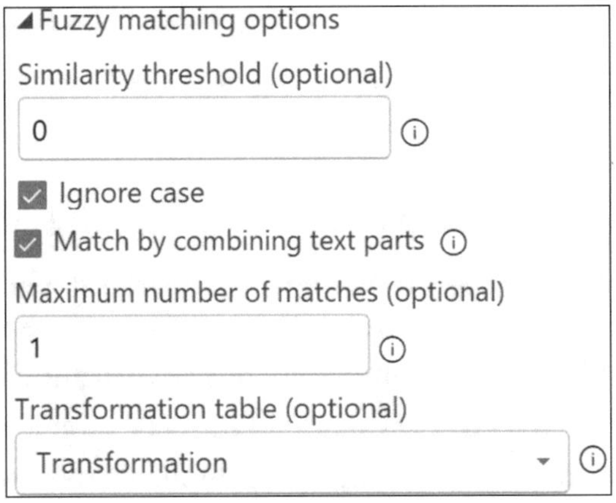

◢ Fuzzy matching options

Similarity threshold (optional)

0 ⓘ

☑ Ignore case
☑ Match by combining text parts ⓘ

Maximum number of matches (optional)

1 ⓘ

Transformation table (optional)

Transformation ▾ ⓘ

Figure 9-53. *The Fuzzy matching options*

The objective of the fuzzy matching was to merge the PRICE column from the Products table into the Sales table so that we could add a custom column for the REVENUE by multiplying the QUANTITY by the PRICE; see Figure 9-54.

1²₃ QUANTITY	ABC PRODUCT	1²₃ PRICE	$ REVENUE
145	Chianti	40	5,800.00
111	Sauvignon Blanc	40	4,440.00
99	Grenache	30	2,970.00
110	Bordeaux	75	8,250.00
291	Grenache	30	8,730.00
233	Pinot Grigio	30	6,990.00
187	Pinot Grigio	30	5,610.00
267	Chianti	40	10,680.00
99	Chardonnay	150	14,850.00
132	Shiraz	78	10,296.00
108	Piesporter	125	13,500.00
273	Champagne	200	54,600.00
90	Chenin Blanc	50	4,500.00
154	Sauvignon Blanc	40	6,160.00

Figure 9-54. *The REVENUE column is the result of merging queries using the fuzzy match*

Therefore, it's fuzzy matching that will allow you to merge columns from different tables without needing an exact match, with the option to set a similarity threshold and a matching tolerance. You can even map rogue values to their counterparts using a transformation table. It's important to note that fuzzy matching is exclusively supported for merge operations involving text columns.

Q62. Can I Bin Values in Power Query?

Power Query is a great tool for generating bins for analyzing values that fall between numeric ranges. Consider Figure 9-55 where we are analyzing QUANTITY values in a column chart by binning them into numeric ranges.

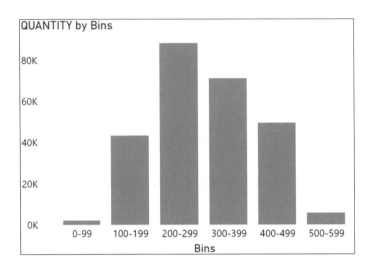

Figure 9-55. *Binning the QUANTITY values into numeric ranges*

Of course, this could be done in Power BI, using DAX in a calculated column, but as we've been advised, calculated columns should be avoided if possible. Besides this, we would have to construct a DAX expression to group the values. If we use Power Query, we can simply provide examples of the bins we need. To do this, select the column whose values you want to bin. In our example, this is the QUANTITY column. Then on the **Add Column** tab, use the dropdown on the **Column From Examples** button, and select **From Selection**; see Figure 9-56.

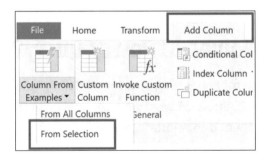

Figure 9-56. *To bin values, use Column From Examples*

A blank column is generated on the right of the Power Query window. In the first row of the column type, the bin range into which that first value will fall. Repeat for another few rows until the column is populated with the correct bins. For example, in Figure 9-57, we had to provide three examples in the first three rows; see Figure 9-57.

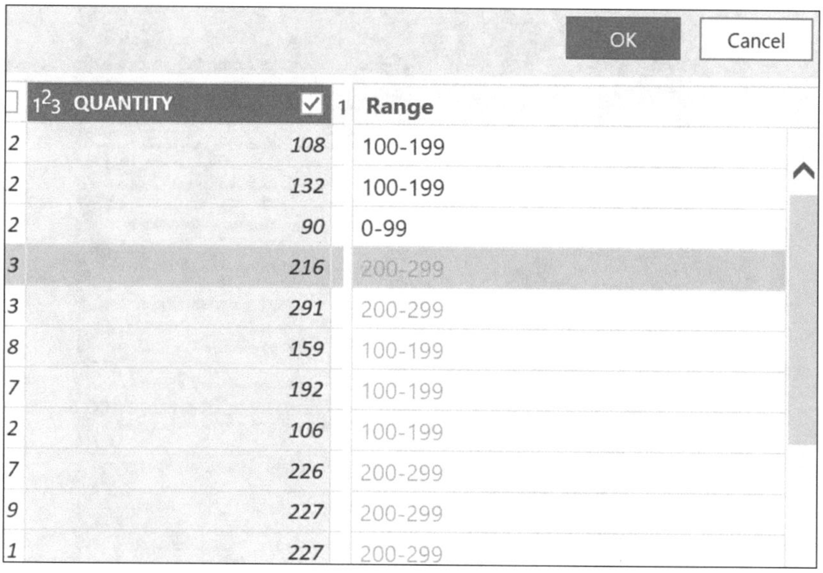

Figure 9-57. Provide examples of the bin ranges that are required for each value

Click **OK** when your examples have generated the values you need. You can then rename the column that has been generated as required.

Q63. Can I Rank Values in Power Query?

Yes, you can, and it's easier than using DAX (see QU48, "How Can I Rank or Find TopN Customers or Products?"). For example, we may have a QUANTITY value for each of our customers (see Figure 9-58) by which we want to rank our customers.

	ᴬᴮC CUSTOMER	1²3 QUANTITY
1	New Castle Ltd	148
2	Cheney & Co	129
3	Melbourne Ltd	200
4	Busan & Co	150
5	Shanghai Ltd	110
6	Victoria Ltd	89
7	Suresnes Ltd	161
8	Harwich Ltd	127
9	Beaverton & Co	92
10	Martinsville Bros	215
11	Nizhny Novgorod & Co	114
12	Burningsuit Ltd	112
13	Kirkland Ltd	217

Figure 9-58. *We want to rank our customers by their quantity*

All we need to do here is to sort on the QUANTITY column and then add an **Index Column** using the button on the **Add Column** tab. We can then rename to column to RANK; see Figure 9-59.

	ABC CUSTOMER	123 QUANTITY	123 RANK
1	Charleston Ltd	246	1
2	Miyagi & Co	243	2
3	Yokohama & Co	240	3
4	Newcastle upon Tyne & Sons	240	4
5	Plattsburgh Ltd	239	5
6	Columbus & Sons	239	6
7	Loveland & Co	234	7
8	St. Leonards Ltd	226	8
9	Brooklyn Ltd	221	9
10	Littleton & Sons	220	10
11	Kirkland Ltd	217	11
12	Martinsville Bros	215	12
13	Warsaw Ltd	214	13

Figure 9-59. *To rank by QUANTITY, sort the column, and then add an Index Column, renamed to RANK*

However, maybe you need to rank your customers within groups. Consider Figure 9-60, where, for example, we want to rank customers by the QUANTITY column, grouped by REGION.

	ABC CUSTOMER	ABC REGION	123 QUANTITY
1	Hervey Bay Bros	Sweden	50
2	Haney Ltd	Sweden	53
3	Snoqualmie & Sons	United States	57
4	Fremont & Sons	Germany	58
5	Parkville Ltd	United States	60
6	East Orange & Co	Sweden	64
7	Lavender Bay Ltd	Germany	66
8	Issaquah & Co	Sweden	70
9	Burlington Ltd	United Kingdom	70
10	Port Orchard & Sons	United States	73
11	Branch Ltd	United Kingdom	74
12	Ballard & Sons	United Arab Emirates	76
13	Fort Worth Ltd	United States	78
14	Barstow Ltd	United Kingdom	84

Figure 9-60. *We want to rank customers by quantity within their regions*

To do this, first sort the QUANTITY column. Then select the REGION column, and on the **Transform** tab, click the **Group By** button. In the **Group By** window, the only option that requires changing is "Operation" where we must select "All Rows"; see Figure 9-61. Click **OK** and a column named "Count" is generated.

Group By

Specify the column to group by and the desired output.

⦿ Basic   ◯ Advanced

REGION ▾

New column name	Operation	Column
Count	All Rows ▾	▾

Figure 9-61. *Select "All Rows" in the "Operation" dropdown*

Now on the **Add Column** tab, we can select **Custom Column**, and type this M Code, as shown in Figure 9-62.

```
Table.AddIndexColumn([Count],"RANK",1,1)
```

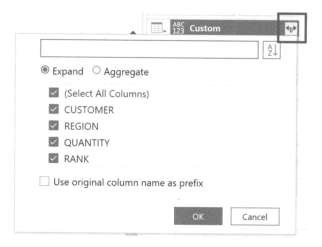

Figure 9-62. *Create a Custom Column with the M Code shown*

The M code adds an index column, called "RANK," that counts each row in the group (because we grouped using "All Rows"), starting with 1 and incrementing by 1. When the Custom column has been added, we can then remove the REGION and Count columns. Now all we must do is expand the Custom column as shown in Figure 9-63.

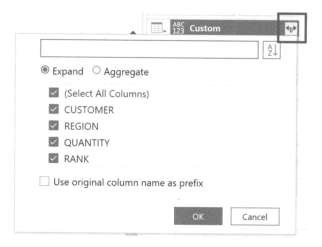

Figure 9-63. *Expand the Custom column*

We now have our customers ranked by quantity, within each region; see Figure 9-64.

	CUSTOMER	REGION	QUANTITY	RANK
1	Hervey Bay Bros	Sweden	50	1
2	Haney Ltd	Sweden	53	2
3	East Orange & Co	Sweden	64	3
4	Issaquah & Co	Sweden	70	4
5	Rhodes Ltd	Sweden	91	5
6	Erlangen & Co	Sweden	103	6
7	Chandler & Sons	Sweden	117	7
8	Lake Oswego & Co	Sweden	143	8
9	Landstuhl Ltd	Sweden	186	9
10	Hawthorne Bros	Sweden	195	10
11	El Cajon & Sons	Sweden	202	11
12	Warsaw Ltd	Sweden	214	12
13	Snoqualmie & Sons	United States	57	1
14	Parkville Ltd	United States	60	2
15	Port Orchard & Sons	United States	73	3
16	Fort Worth Ltd	United States	78	4

Figure 9-64.   *The customers ranked in each region*

So it really is quite straightforward to rank values using Power Query and even to rank by grouped rows.

Q64. How Can I Calculate Cumulative Values Using Power Query?

In Q27 "How Do I Calculate Cumulative Totals?" we explored how we could use DAX in Power BI Desktop to do this. If you want to use Power Query, this is a little trickier. For example, in Figure 9-65, using Power Query we have calculated the cumulative totals for the QUANTITY column.

	SALEDATE		1²3 QUANTITY		1²3 Cumulative	
1	01/01/2021		108		108	
2	02/01/2021		206		314	
3	03/01/2021		239		553	
4	04/01/2021		136		689	
5	05/01/2021		106		795	
6	06/01/2021		117		912	
7	07/01/2021		140		1052	
8	08/01/2021		140		1192	
9	09/01/2021		159		1351	
10	10/01/2021		252		1603	
11	11/01/2021		106		1709	

Figure 9-65. *Cumulative values for the QUANTITY column*

There is no generic way to calculate cumulative values using Power Query, so you must resort to using M code instead. To generate the Cumulative column as shown in Figure 9-65, first sort the rows by the SALE DATE column. Then using the **Index Column** button on the **Add Column** tab, add an index column, ensuring it starts at "1." To calculate the cumulative values, add a **Custom Column**, with the following M code entered into the formula bar (see Figure 9-66):

```
List.Sum(List.Range(#"Added Index"[QUANTITY], 0, [Index]))
```

Custom Column

Add a column that is computed from the other columns.

New column name

Cumulative

Custom column formula ⓘ

```
= List.Sum(List.Range(#"Added Index"[QUANTITY], 0, [Index]))
```

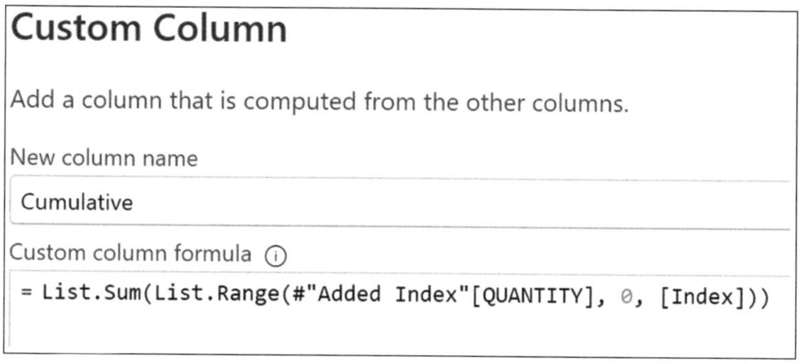

Figure 9-66. *The M code for the Cumulative column*

This function builds a list of values from the QUANTITY column, starting at **0** and incrementing by the values in the Index column. It does this for every row. The List.Sum function then sums the values in this list. The reference to the "Added Index" step is simply a reference to the previous step.

However, the issue now becomes how to make the query run faster because generating these cumulative totals will render the query very slow, even with a modest number of rows. The problem lies in the calculation itself; if we look more closely at the List.Range function (see Figure 9-67), we can see that for every evaluation, the list of values to be summed (by the List.Sum function) is regenerated in memory, and therefore this virtual list grows with each evaluation.

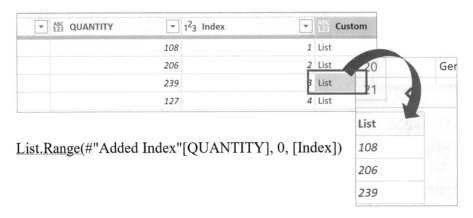

Figure 9-67. *Using the List.Range function, the list of values is regenerated for every evaluation*

What we can do here is speed up the query by inserting a step prior to adding the cumulative column. Using the List.Buffer function, this additional step will generate in memory a once-only list of values to be accumulated so that it can then be referenced by the List.Range function.

To do this, select the "Added Index" step in the APPLIED STEPS pane, and then click on the "**fx**" button to the left of the formula bar. After confirming that you want to insert a step, type this formula:

```
= List.Buffer (#"Added Index"[QUANTITY])
```

It makes sense to then rename this step from "Custom1" to something more meaningful such as "TempList."

Now, all that remains is to edit the "Added Custom" step that contains the List.Range function. Use the formula bar to do this; see Figure 9-68. This will be the M code in the formula bar:

```
= Table.AddColumn(TempList, "Cumulative",
each List.Sum(List.Range(TempList[QUANTITY], 0, [Index])))
```

Edit the M code to this:

```
= Table.AddColumn(#"Added Index", "Cumulative",
each List.Sum(List.Range(TempList, 0, [Index])))
```

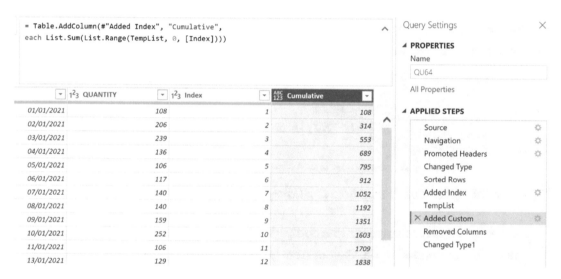

Figure 9-68. *Edit the "Added Custom" step in the formula bar*

We can use Power Query to generate cumulative values, but problems can arise from longer refresh times if you do so. For this reason, you may prefer to use DAX in Power BI instead.

Q65. How Can I Make My Queries More Flexible?

You can use query parameters to overcome the problems of hardcoding values into your queries and therefore make them more flexible. For example, using parameters you can substitute values on which you want to filter your data. However, using parameters can also extend to overcoming issues with hardcoding file paths, web addresses, and server names into connection steps. Using parameters means that you can dynamically switch between different values, enabling greater flexibility in how you want to query your data.

If you feel parameters would be beneficial to your queries, the first step is to determine how you're going to create new parameters. If you'd like the ability to create a parameter wherever it's possible to do so, you must check on the "Always allow parameterization" option which you will find on the **File** tab, **Option and settings** and **Options**, under the **Power Query Editor** category; see Figure 9-69.

Figure 9-69. *Check on the option to allow parameterization*

Alternatively, for the current report only, you can use the **View** tab in Power Query, and check "Always allow" in the **Parameters** group; see Figure 9-70.

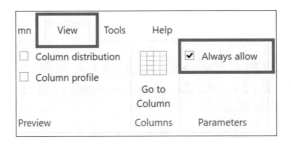

Figure 9-70. *You can always allow parameters for the current report*

If you have allowed parametrization, wherever it's possible to use values from parameters, you will see a dropdown button allowing you to create a parameter; see Figure 9-71.

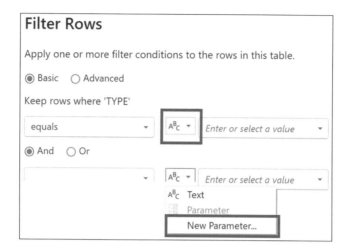

Figure 9-71. *You can add new parameters from the dropdown button*

However, whether you have allowed parameterization or not, you will always have the option to create new parameters and manage existing parameters from the **Manage Parameters** button on the **Home** tab; see Figure 9-72.

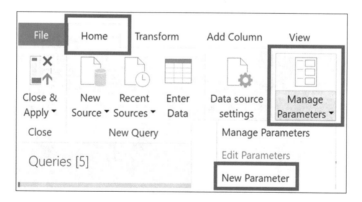

Figure 9-72. *You can always create parameters from the Manage Parameters dropdown on the Home tab*

Now we know where to find the option to create parameters, let's explore three examples of how parameters can help you make your queries more flexible. In the first of these examples, we will take a simple filter such as the one you can see applied to the REGION column in Figure 9-73 where we have hardcoding "France" as the value to filter. Instead of the hardcoded value, we would like to dynamically select different regions on which to filter the data.

⨯ ✓ _fx_	= Table.SelectRows(#"Removed Columns2", each [REGION] = "France")				
▥. ▦ SALE DATE	▾	A^B_C CUSTOMER ▾	A^B_C REGION ▼	1²3 QUANTITY	▾
1	03/06/2021	Nizhny Novgorod & Co	France	114	
2	07/06/2021	Chatou & Co	France	162	
3	08/06/2021	Milsons Point Ltd	France	153	
4	08/06/2021	Kirkland Ltd	France	217	
5	12/06/2021	Kirkland Ltd	France	157	
6	13/06/2021	Kirkland Ltd	France	159	
7	14/06/2021	Nizhny Novgorod & Co	France	239	
8	15/06/2021	Burningsuit Ltd	France	112	
9	16/06/2021	Kirkland Ltd	France	214	

Figure 9-73. *A simple filter on the REGION column, filtering "France"*

Because we have many regions by which we can filter the data, we can use Power Query to generate a list of regions for us. To do this, right-click the REGION column, and select "**Add as New Query**" from the dropdown menu. A new list query, named REGION, is added to the Queries pane. On the **Transform** tab of the list query, select to **Remove Duplicates** to edit the list to unique values; see Figure 9-74.

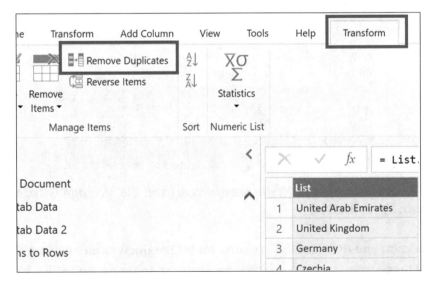

Figure 9-74. *Remove the duplicates from the list query*

Now we can open the Manage Parameters window as shown in Figure 9-72 and create our parameter. We have named it "Select Region," and you can see in Figure 9-75 how the other options have been completed in the Manage Parameters window. We selected "Query" in "Suggested Values," and "REGION" is the list query selected in the "Query" dropdown.

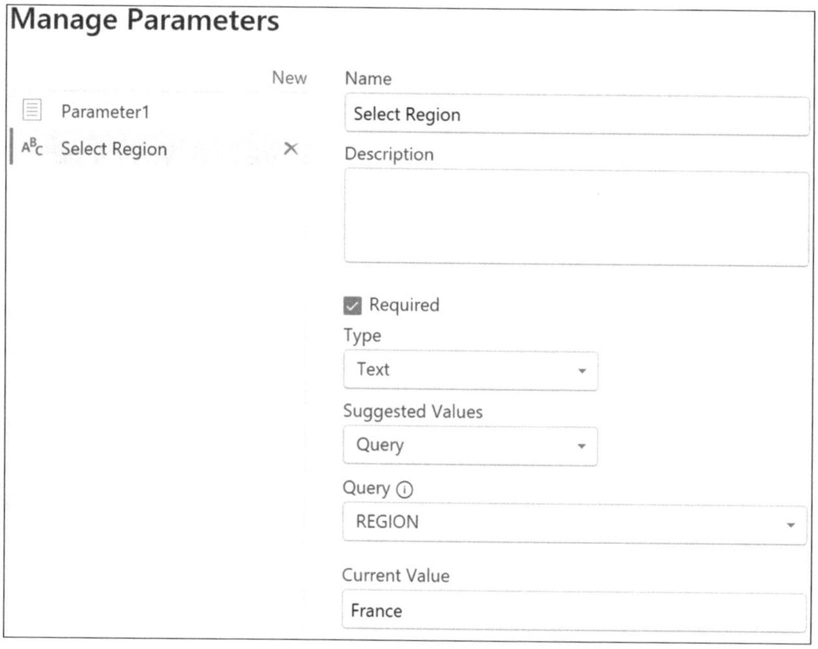

Figure 9-75. *Creating the "Select Region" parameter*

When you click OK in the Manage Parameters window, a parameter query is added to the Queries pane, named "Select Region (France)."

To apply the "Select Region" parameter to the filter, on the filter dropdown of the REGION column, select "Text Filters" and then "Equals." Now use the parameter dropdown in the Filter Rows dialog as shown in Figure 9-76.

Filter Rows

Apply one or more filter conditions to the rows in this table.

◉ Basic   ○ Advanced

Keep rows where 'REGION'

| equals | ▾ | 🗒 ▾ | Select Region | ▾ |

Figure 9-76. *Apply the parameter by using the parameters dropdown*

To change the parameter value, select the parameter query in the Queries pane, and edit accordingly; see Figure 9-77.

Current Value

Germany

Manage Parameter

Figure 9-77. *Changing the parameter value*

However, rather than switching parameter values in Power Query, you may prefer to do this in Power BI Desktop. If so, use the dropdown on the **Transform** data button on the **Home** tab, and select **Edit parameters**, Figure 9-78.

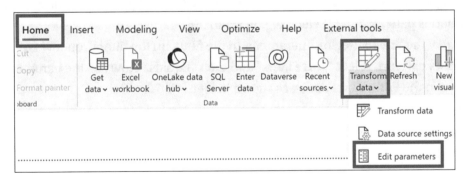

Figure 9-78. *You can edit the parameter values in Power BI Desktop*

Using parameters in this way can become particularly useful if you are using Power BI templates which can be created by selecting **Export** from the **File** tab. When you create a new report using a Power BI template, the Edit parameters dialog shows when the template is opened.

In our second example, we have created a parameter named "TOPN," this time using "List of values" in the "Suggested Values" dropdown and typing the topN values to be used. This parameter was then applied to the "Keep Top Rows" transformation (on the **Keep Rows** button); see Figure 9-79.

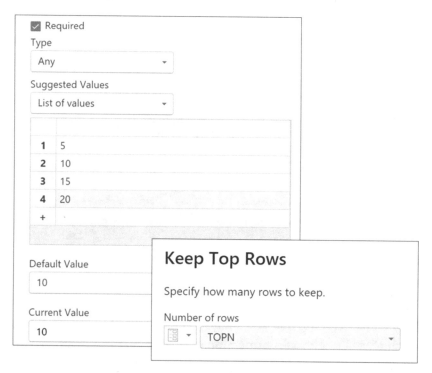

Figure 9-79. *The "TOPN" parameter for the Keep Top Rows transformation*

In our last example, we're going to explore how we can use parameters to dynamically switch folders from where the source data is imported. Consider Figure 9-80 showing three folders, "Sales 2022," "Sales 2021," and "Sales 2020," saved on the "C" drive, each containing monthly data for "Jan," "Feb," and "Mar" in three Excel files.

Figure 9-80. *Three folders each with three Excel files saved on the "C" drive*

We can create a parameter that can be used by the "Source" step whereby we can change which folder and file will be imported. To do this, create two parameter queries, "Select Year" and "Select Month." These will both contain lists of values as shown in Figure 9-81.

	Select Year		Select Month
Suggested Values		Suggested Values	
List of values ▾		List of values ▾	
1	Sales 2020	1	\Jan
2	Sales 2021	2	\Feb
3	Sales 2022	3	\Mar
+		+	

Figure 9-81. *The parameter values for "Select Year" and "Select Month"*

Now we must import any one of the Excel files from any one of the folders. Once we have done this, we can edit the "Source" step by clicking on the gear symbol to the right of the step. In the Excel Workbook connection dialog, using the **Advanced** options, we must now dissect each element of the path name, substituting the parameters where applicable. You can do this by using the "**Add part**" button and selecting the parameter dropdowns, as shown in Figure 9-82.

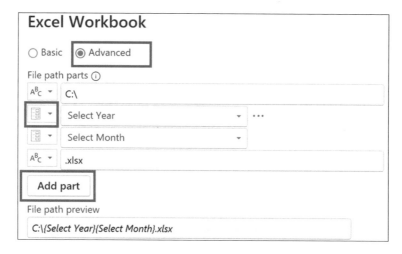

Figure 9-82. *Edit the Source step to include the parameter values*

Now we can switch these parameter values as required by selecting the parameter queries in the Queries pane and editing accordingly.

Therefore using parameters in this way can relieve the frustrations of hardcoding values into the queries themselves and manually editing steps or the M code. We have only scratched the surface of possibilities here, and I'm sure you will find many more uses of parameters when you design your queries.

This question also completes The Power BI Compendium. I hope you have enjoyed and benefited from browsing the answers to the questions and can now achieve more accurate and powerful insights into your data.

Index

© Alison Box 2023
A. Box, *A Power BI Compendium*, https://doi.org/10.1007/978-1-4842-9765-0

Printed in the United States
by Baker & Taylor Publisher Services